Gilbert Simondon

Gilbert Simondon

Information, Technology and Media

Simon Mills

ROWMAN &
LITTLEFIELD
INTERNATIONAL
London • New York

Published by Rowman & Littlefield International, Ltd.
Unit A, Whitacre Mews, 26-34 Stannary Street, London SE11 4AB
www.rowmaninternational.com

Rowman & Littlefield International, Ltd. is an affiliate of Rowman & Littlefield
4501 Forbes Boulevard, Suite 200, Lanham, Maryland 20706, USA
With additional offices in Boulder, New York, Toronto (Canada), and London (UK)
www.rowman.com

British Library Cataloguing in Publication Information Available
A catalogue record for this book is available from the British Library

ISBN: HB 978-1-7834-8148-4
ISBN: PB 978-1-7834-8149-1

Library of Congress Cataloging-in-Publication Data

Names: Mills, Simon (Senior Lecturer), author.
Title: Gilbert Simondon : information, technology, and media / Simon Mills.
Description: Lanham : Rowman & Littlefield International, 2016. | Includes bibliographical references and index.
Identifiers: LCCN 2015047674 (print) | LCCN 2016001776 (ebook) | ISBN 9781783481484 (cloth : alk. paper) | ISBN 9781783481491 (pbk. : alk. paper) | ISBN 9781783481507 (Electronic)
Subjects: LCSH: Simondon, Gilbert. | Individuation (Philosophy) | Technology--Philosophy.
Classification: LCC B2430.S554 M55 2016 (print) | LCC B2430.S554 (ebook) | DDC 194--dc23
LC record available at http://lccn.loc.gov/2015047674

∞™ The paper used in this publication meets the minimum requirements of American National Standard for Information Sciences Permanence of Paper for Printed Library Materials, ANSI/NISO Z39.48-1992.

Printed in the United States of America

For Sonal

Contents

Acknowledgements

There are many people whose support and advice have directly influenced the production of this book and without whom it wouldn't have been possible.

I am indebted to Iain Grant, my doctoral supervisor, whose indefatigable enthusiasm, coupled with a seemingly inexhaustible philosophical knowledge, provided the overwhelming impetus for the development of this book.

I must thank my friends and colleagues in the Media and Communication Department at DMU, in particular Stuart Hanson, Alastair Gordon, Paul Smith and Jim Russell, for their ongoing support and putting up with my too-frequent updates. Dave Everitt also deserves a special mention for his friendship and wisdom. Additionally, Helen Wood was a great support during the earlier stages of this project, for which I remain appreciative.

This text was also enriched by my good fortune to have met some notable fellow Simondon researchers. In particular I would like to mention Anne Lefebvre and Cecile Malpasina for sharing their fascinating ideas and knowledge. I must also acknowledge Dan and Nandita Mellamphy for sharing translations of Simondon's texts and Mark Hansen for supplying me with a then-unpublished article, which was very useful towards the end of the project.

I would also like to thank my parents for all their help and support over the years.

I also want to mention the late Gregory McCullough, my memories of whom are still inspirational all these years later, and also Randy Adams, who will always be the dude.

Finally, but by no means least, I must thank my partner Sonal Patel whose love, support and encouragement, as well as considerable tolerance for my long working hours, was vital for this book's completion. I cannot thank her enough for the sacrifices she made so that I could pursue my interest.

Introduction

It is appropriate that the work of Gilbert Simondon is finally becoming more widely known at this particular moment, given that we find ourselves in a situation in which new technologies once again challenge cultural values. That this is occurring alongside a reassessment regarding what of use might be taken from modernity, often through an attempted retrieval or adaptation of visions of the future imagined in the sixties, only adds to its relevance. Although his main works were written in a period stretching between 1952 and 1966,[1] Simondon's oeuvre is still imbued with an untimeliness, simultaneously of its era, yet also still prescient for us today. In this he resembles the figure of the inventor who he holds in such high esteem, whose engagement with the material of the present brings forth an unforeseen and unforeseeable future.

One possible reason for this untimeliness is due to the vision we find in his work of a modernity (albeit in crisis) still replete with the potential for *progress*. It's not a word that is easy to use this side of the post-modern moment (especially for those on the political left), reminding us as it does of the spectre of western colonialism and dehumanizing instrumental rationality. However, as with so much in Simondon's work, he reimagines the meaning of concepts so they escape their old substantiality.

Writing when he was, the crisis in question was that of a state of alienation between human culture and industrial technology, specifically regarding the humiliation of technology in its role in demeaning forms of labour. This is a perspective that shares plenty with other diagnosticians of industrial modernity, such as Heidegger, Adorno and Marx. Although alive to the cataclysmic potential of industrial technology for the native state of nature on the planet, as well as for society, for Simondon it wasn't an option to call time on the future. Invention cannot and will not cease and nor can culture attain a stable homeostatic equilibrium (despite the well-intentioned claims of cyberneticians such as Norbert Wiener and Stafford Beer) in its relations to technology; to do so is to invite entropy.

It is indicative of the strength of his vision that, unlike many, Simondon didn't see in the cataclysmic events of the twentieth century the need to dispense entirely with the idea of progress, although how such progress should be envisioned required reworking. This vision is not founded on some blind faith in reason; its source is found in a deep appreciation, knowledge and wonder at the role technologies and reflexive thought play in our world and evolution.

For Simondon the problem of industrial modernity was due to an extreme imbalance between the phases of technology and culture, which led to a dangerous technocratic attitude. What was required was the rebalancing of cultural values in response to the new technological reality they now operated within. This didn't mean the imposition of pre-existing humanist values upon technological inventions as those in the Frankfurt school proposed, but rather the need to understand how the dynamic interplay of technological development with culture was productive of new values and desires that required acknowledgement.

If human culture is to stay in phase with its technical inventions what is also required is an appropriate updating of its system of knowledge. Simondon saw the need for a change in the operation of knowledge from that which was based on a hylemorphic notion of the individual, which has pervaded thought since Aristotle, to one founded on an ontogenetic account of individuation. Buoyed by developments in the new science of information, as conceived by Norbert Wiener and the cyberneticists, Simondon proposed an axiomization of the various domains of knowledge leading to a revised modernity founded not on universal values but on a consistency of understanding bringing forth a unity.

Undoubtedly Simondon was attracted to the interdisciplinary cyberneticist approach to knowledge as well as its informational foundation, which permeated all domains. However, the cybernetic account of information as developed by Shannon and Wiener is found to be inadequate to the task, being still haunted by a residual hylemorphism. Additionally, the cyberneticists' use of analogy also lacks a suitable sensitivity to the role of operations.

In response Simondon proposes a notion of information that is reformulated to give an ontogenetic account of the development of individuals across three regimes of nature: the physical, the vital and the psychosocial. It is impossible to understand Simondon's philosophy of technology without first comprehending this revision to the notion of information and how this enables him to develop a new philosophy of nature that also entails a reworking of knowledge in all domains. As such what Simondon proposed was a renewal of the encyclopaedic project of the enlightenment involving a unification of separate domains of knowledge by revising the perspective on their subject to that of the ontogenetic operation of individuation.

In an essay from 1960 published later in *L'individuation à la lumière des notions de forme et d'information* (2013), Simondon proposes the use of his transductive theory of individuation as an axiomatic method to create 'a general theory of the human sciences [sociology] and psychology' (Simondon, 2013: 537). Such a universalizing theory would regard the whole of human reality, in both its social and psychological dimensions, as comprehensible in the same individuating movement. Such a theory, underpinned by Simondon's notion of information and energetics, sounds radical even today, when there are still many in interdisciplinary enterprises, such as media studies, who still attempt to bind together disparate psychological theories with those from sociology in the hope of a coherent outcome. One aim I have for this volume is that it might signal a possible way forward, especially given the radically environmental nature of our contemporary media technologies.

Another important aspect of Simondon's contemporary relevance is his humanism. As with modernism, the term *humanism* is currently going through a process of renegotiation as evidenced by the contemporary interest in theories of post-humanity and the inhuman. Simondon was also ahead of his time in this respect, with his call for a new humanism based on twin requirements: to be done with both the notion of a human essence as well as the pervasive understanding that technology is purely instrumental and not a true part of culture. It was these mistaken notions that Simondon argues helped forge the form of alienation that plagued the time in which he worked. His contention is that we can never be done with alienation, that each age has to contend with a new form of it as a consequence of developments in technology. As such humanism requires constant reinvention in response to these new modes of alienation, thereby rebalancing the phases of culture and technicity.

It's possible that Simondon may have had some sympathy for Kittler's prognostication to be done with 'so-called Man', but only to the extent that this is through a transformation, an inventive impulse and not a subsumption into some kind of digital entropy. His insistence that there is no essential human characteristic upon which an anthropology can be built entails that like the human, humanism is also always a work in progress in congress with technology.

I was once told by another media academic that Simondon had nothing to offer the study of media. This was an attitude I found both surprising and blinkered. In turn, I've always found it fascinating that in much of the theoretical work of British media studies, as well as that of critical theory, so little attention has been paid to the role of technology in the constitution of the media in question. It is as if that which is worth commenting on regarding media is in fact nothing technological, to abuse a phrase. In media and communication theory the result of the disagreement of Williams with what he understood as Mcluhan's blunt techno determinism led to a culturalist attitude that never adequately grasped the place of technology. What I find compelling in Simondon is how he helps us ontogenetically track the development of technology, from tool to instrument, individual to ensemble and from network to environment, in relation to biological, psychic and sociocultural progression. The environmental aspect of contemporary media is too extensive to be adequately understood through just discourse, representation or the values of critical theory. These have their place but must be situated in a far broader vision. As the media theorist John Durham Peters writes, 'The time is ripe for a philosophy of media. And a philosophy of media needs a philosophy of nature. Media are not only devices of information; they are also agencies of order' (Peters, 2015: 1). I couldn't agree more, and Simondon provides just such a philosophy, replete with an ontologically consistent way of understanding what environmentality means, and the radical openness it entails. Another aspect of Peters' work is his interest in the importance of technical infrastructure, which also mirrors Simondon's concerns in thinking about the meaning of networks.

To claim to study media without an understanding of media technology or the range of levels with which they interact with individuals, collectives and the wider environment seems a partial enterprise. It is as if we were to understand the Pantheon in Rome just through its symbolism and meaning without recourse to a discussion of the history of the techniques required for its building. Of course some might argue that what is essential is the political and social structure that led to its being built. That it is cultural first and foremost. However, without the expert technical knowledge and its application to the production of the various types and density of bricks required for its construction,[2] the iconic dome and oculus, those wonderful symbols of the mediation between earth and the heavens, would never have been possible. Further, the infrastructure upon which that great civilization arose and relied upon would also have been impossible. The roads, aqueducts and sewers were all constructed due to this technical knowledge and use of materials. This may seem obvious, but it is just such a blindness that infects many discussing media today.

The rationale of this book is to explore Simondon's axiomatic theory of ontogenesis in order to understand how it is able to produce a coherent philosophical project that includes an ethics, epistemology, aesthetics and technical philosophy that is inclusive of the various material operations that are involved across disparate domains. Given such a realist ontology this also entails an investigation into the nature of causality such that these domains can cohere together.

The book is split into two main parts. In the first four chapters I offer an overview and introduction to Simondon's philosophy of individuation. In chapter 1, prior to describing Simondon's positive theory of individuation, I explain what he finds lacking in traditional theories of form. This includes the problems Simondon has with theories such as hylemorphism and atomism. Given my own focus on media technology I pay particular attention here to the cybernetic notion of information that both inspired Simondon, yet that he also found lacking. Given this theory plays such a central role in the subsequent establishment of media studies, gaining a fuller understanding of it and of Simondon's issues with it is necessary. Simondon's core theory of operation and structure, called *allagmatics*, is also introduced in this chapter, which involves a critical investigation of the use of analogical reasoning.

An account of the key components of Simondon's own ontology begins in chapter 2 by elucidating his theory of the transductive individuation of physical being. This involves explaining how Simondon builds on cybernetics by transforming the notion of information and coupling it with the thermodynamic concept of the phase-shift. Additionally, Simondon's account of individuation will be contrasted to some contemporary explanations of phenomena given in complexity theory in order to help demonstrate the coherence of his account of causality.

The capability of Simondon's allagmatics to think causality through the interplay of operation and structure is evident in the importance that topology plays in his account of vital individuation in chapter 3. Some other core Simondonian concepts are also explored such as order-of-magnitudes and levels that further deepens Simondon's accounts of ontogenesis. In doing this the similarities Simondon shares with the biologist Stuart Kauffman's work will also be explored as well as how Simondon provides a way to critique some of the residual Aristotelian tendencies therein.

In chapter 4 the development of Simondon's ontology progresses with two theories that are crucial for understanding his later work on technology: that of psychic and collective individuation and the theory of the image-cycle. What Simondon's description makes apparent is that there is a gradation of development from the sensual towards perceptual consciousness, which indicates the possibility for a politics of sensibility, which could partially undermine or at least complement the politics of intelligence proposed by Bernard Stiegler. That is to say that Simondon considers the gradations of sensual and affective experience, which condition phenomenological consciousness, rather than concentrating predominantly on the operation of an already fully constituted phenomenologizing subject. As such Simondon's work is relevant for those working on theories of affect.

For Simondon the individuation of the psychic is inseparable from that of the collective, and due to the way the modality of collective individuation can be affected by technology this becomes another core part of Simondon's philosophy of technology. It is in this respect that Simondon's phasic notion of social individuation and the transindividual becomes important.

One of the fundamental features of Simondon's conception of individuation is the importance it places on invention. As described in chapter 2, transduction, as an informational theorization of the phase-shift, describes a productive process that involves leaps between levels of being. It is this axiomatic application of transduction to all levels of being that informs Simondon's account of the progressive development of the imagination as a site for epistemo-genesis in the organism through a recursive process culminating in exteriorization as the invention of an image-object. It is important that this process is not only productive of technical objects (and other artefacts), but also and reciprocally technical schemas of thought by which the world is understood.

Having now detailed the comprehensive scope of Simondon's ontology, chapter 5 involves an exploration of his philosophy of technology. As stated at the beginning of *On the Mode of Existence of Technical Objects* the main concern of that study is to address the problem of culture being out-of-step with technology, thereby creating a situation of alienation. In this chapter I contrast Simondon's account with other significant work in the area, such as Heidegger and Marx. What is found is that in addressing this problem a tension is developed in Simondon's work between the productive and inventive causality of individuation and the importance placed on regulative values such as those of magical-unity and techno-aesthetics. The resolution to this tension will be found in the nature of regulative causality that Simondon develops from cybernetics and that is described in his account of the relation between culture and technics. This is explored in chapter 6.

The aim of chapter 7 is to outline some of the possible ways that Simondon is useful for thinking about the contemporary technological situation, especially networked digital technology. In this chapter Simondon is contrasted with two significant contemporary thinkers of technology, Bruno Latour and Bernard Stiegler. Latour is relevant due to his intentions regarding ontology, although I will argue that he remains too much of an anthropologist to fulfil his promise of restoring technologies ontological dignity. As a thinker who is significantly influenced by Simondon's work and whose focus is also on technology, Bernard Stiegler is of obvious interest. However, like Latour, we also find his work too anthropological and his account of technology as tertiary retention too limited.

Chapter 8 looks at some of the ways that Simondon's thought can be applied to contemporary media technology. It is my contention that with his philosophy of individuation Simondon's work is well placed to help overcome an explicit culturalism that still inheres in much work theorizing media and communication. Given the environmental scope of contemporary media technology, as well as its modes of operations across multiple ontological levels and temporalities, I argue that a more balanced perspective on the relation of culture with technology is required. To explore this possibility Simondon's work is first contrasted with that of British cultural studies and then Friedrich Kittler before pointing to some possible ways to progress.

Given that this book predominantly approaches Simondon's work through his relationship with cybernetics, the final chapter attempts to redress the balance and describes some of the other important influences on his work and situate it in a more philosophical context. In particular his relation to Merleau-Ponty and the broader current of French scientific epistemology is described. In his fascinating text on the history of AI and cybernetics, *On the Origins of Cognitive Science: The Mechanization of Mind*, Jean Pierre-Dupuy makes the claim that 'the encounter between first-order cybernetics and phenomenology had been missed' (2009: 102). One of the things I hope to have demonstrated is that this opportunity wasn't missed but has been fruitfully undertaken by Simondon, even if that meant that both cybernetics and phenomenology were transformed in the process. Additionally, in this chapter, the coherence of Simondon's method as an account of causality is explored by contrasting it with some recent analytic work in that area.

Like Simondon's project this text covers a lot of ground. Although my overall aim is to consider how we can think about media technology using Simondon, it should be clear we can only do so through understanding his philosophy of nature. I'm aware that this approach runs the risk of trying to please two different audiences, those with a purely philosophical interest in Simondon's work and those interested in contemporary media. I hope I've managed to do justice to Simondon's ecumenical vision in this text to persuade readers that these two concerns are productively related.

NOTES

1. Although Simondon taught until 1983, prior to an illness and an early death in 1989, his main works were written earlier in his career. Despite this, much of his work wasn't published until after his death. His main doctoral thesis *L'individuation à la lumière des notions de forme et d'information* wasn't published in its entirety in one volume until 2005. Prior to this it had been published separately in two parts, the first part *L'individu et sa genèse physico-biologique* in 1964, and the second *L'individuation psychique et collective* in 1989. The classic work on technology, *Du mode d'existence des objets techniques*, which was Simondon's complementary thesis, was initially published in 1958. Since the publication of his complete thesis in 2005 and the subsequent interest in his work, a number of other volumes of essays and courses have been published.

2. The dome is made possible, in part, due to the cement bricks of varying densities, the lightest made possible by the variation of mixing of light tufa with scoria. The level of knowledge and ability required both to develop this level of technique for producing infrastructural materials as well as to push the limitations of their use is easy to underestimate when discussing Roman culture.

Chapter One

Precursor to a Theory of Individuation

Given the encyclopaedic ambition of Gilbert Simondon's oeuvre it is possible to approach it from a number of directions. Until recently, in Anglophone philosophy, it has mainly been encountered through the work on technology or for the influence it had on the ontology of Gilles Deleuze. Less well known is the importance of his work for other subjects such as biology, epistemology and the imagination.

It is with his philosophy of nature, which crystallizes around the notion of individuation, that one can best grasp the fundamental ideas underpinning his work. As Deleuze observed in his 1966 review of Simondon's major thesis, *L'individuation á la lumiére des notions de forme et d'information*, it is a 'profoundly original theory of individuation implying a whole philosophy' (Deleuze, 2004: 86).

This originality is the outcome of Simondon reimagining a number of philosophical concepts with those from other disciplines such as physics, biology and cybernetics in order to reconceive the operational reality of nature. In particular, Simondon's aim was to reject any claim for understanding being as a grounding *substance* that would give entities a claim for a stable and self-contained identity.

In order to understand this we must first understand Simondon's opposition to substantialist philosophical theories that prioritize the ontologically complete individual above the process of their individuation. Taking inspiration from Jung, Simondon compares his own philosophical project to the *Magnum Opus*, the alchemical process whereby base matter is transmuted into the Philosopher's Stone. In his case Simondon proposes the dissolution of traditional philosophical theories of form and individuality, thus clearing the ground for establishing his own theory of individuation:

> The *Magnum Opus* started by dissolving everything in mercury or reducing everything to the state of carbon—where nothing more was distinguished, substances lost their limit and their individuality, their isolation; after this crisis and sacrifice came a new differentiation; it is the *Albefactio* and *Cauda Pavonis*, which makes objects emerge from the confused night, like the dawn that distinguishes them by their colour. Jung discovered, in the aspiration of Alchemists, the translation of the operation of individuation, and of all forms of sacrifice, which presupposes a return to a state comparable to that of birth, that is to say, a return to a state rich with potential, not yet determined, a domain for the new propagation of Life. (Simondon, 2013: 557)

As we shall see in the next chapter, the state that is *rich with potential* for Simondon is that of the metastable pre-individual—it is from this that he will construct 'a new theory of form'. But before doing this, previous theories of form must be submitted to his alchemist's retort. The two main theories that Simondon views as in need of overturning are those of hylemorphism and atomism. In his critique of these theories Simondon contends that the idea of individuation has been undertaken 'either before or after the individuation has taken place, according to whether the model of the individual being used is a physical one (as in substantialist atomism) or a technological and vital one (as in hylemorphic theory)' (Simondon, 1992: 299).

By this he means that these traditional principles of individuation fail to explain how individuation occurs because by beginning their accounts from the stipulation of the already constituted individual, the assumption is it already has. Subsequently, the actual process of individuation remains 'as something that needs to be explained, rather than as something in which the explanation is found' (Simondon, 1992: 299). As Simondon notes, this approach to describing individuation is actually individuation in reverse as it '*accords an ontological privilege to the already constituted individual*' (Simondon, 1992: 298) rather than starting with the process of individuation that, if taken into account, would change our understanding of the nature of that individual. It is the philosophical tradition that accords ontological privilege to individuality, which Simondon sees as mistaken and in need of transformation.

Simondon argues that neither substantialist atomism nor hylemorphism[1] offer suitable accounts of individuation because neither gives a description of the actual processes involved the attainment of form. Rather, each affirms a principle that has been established prior to and separate from any actual process of individuation.

Substantialist atomism suggests a principle of individuation founded on the pre-existence of already individuated atoms brought together by 'cohesive forces' to create 'complex individuals'. As such, Simondon argues, the actual process of the individuation of these individuals and the actual manner of their organization is itself left unexplained.

Another traditional philosophical theory of form is that of hylemorphism, which Simondon pays more attention to given its widespread use for accounting for form in multiple domains (for example, technical, vital, social).

As a theory hylemorphism has been in philosophical use since Aristotle in whose account for something to be a thing it must be composed of both matter and form, the combination of which constitutes the substance of that thing. Although the formal cause is understood to be the more important cause, as it is that which necessitates the kind of substance that the matter will become, the account remains hylemorphic as both form and matter are necessary. The simplest example of hylemorphism is probably that of the use of a mould shaping matter.

Simondon criticizes the hylemorphic account because rather than being located in the individuation process 'the principle is thought to be contained either in the matter or the form, because the actual process of individuation is not thought to be capable of *furnishing* the principle itself, but simply of *putting it into effect*' (Simondon, 1992: 299). Again, a principle of individuation is asserted prior to the actual process itself and the principle tells us nothing about the actual operation of the individuation process in itself.

For Simondon hylemorphism is a technical schema that has been poorly applied by analogy, thereby masking other types of operation. As he makes clear, the technical origin of the scheme does not invalidate it as long as 'the operation which is the basis for the formation of the concepts used passes and is expressed entirely without alteration in the abstract scheme' (1995: 28). An example of an invalid application of this schema Simondon uses is that of the individuation of a brick. The hylemorphic account of the formation of a brick by the action of a mould on clay misses key aspects of the individuation process. In particular the account misses the preparatory aspects of the process, such as the manufacturing and preparation of the mould itself (for example, coating its surface to prevent clay sticking to it), the preparation of the clay so it is ready for moulding as well as the subsequent drying process. In short, Simondon points out the many processes of both micro- and macrophysical mediation that must occur throughout the entire brick-making process. He thereby claims 'the real dynamism of the operation is very distant from being represented by the form-matter pairing' (Simondon, 1995: 28). As such, the operation that forms the basis for the hylemorphic schema is not 'expressed entirely' as 'the form and the matter of the hylemorphic scheme are abstract form and matter' (Simondon, 1995: 28), which are insufficient to account for the 'real dynamism' of individuation.

Form and matter are thus understood as abstractions of the extreme poles of the actual operations that occur during individuation. It is between these theoretical abstractions that Simondon identifies an 'obscure zone' where the *actual* operation of individuation takes place and remains unexplained:

> There's a hole in the hylemorphic representation, making the true mediation disappear, the operation itself which connects one to the other of the two half-channels in instituting an energy system, a state that evolves and must exist effectively for an object to appear in its haecceity. (Simondon, 1995: 40)

The presence of this gap in the hylemorphic explanation necessitates the need for a reformulation of the notion of taking form, one that Simondon achieves by reconceiving it not as a moulding operation but as one of the modulation of an energetic system through which form emerges. The over-abstraction and simplification hylemorphism performs on the operation and matter involved in the attainment of form, both natural and technical, also has subsequent cultural implications leading to a diminishment in understanding and appreciation:

> The cultural schema of opposition between matter and form, which supposes matter's passivity, is extremely impoverished when faced with the valorization of matter that results from technical operations; matter harbours functional characteristics corresponding to cognitive schemas and axiological categories that culture cannot offer. (Simondon, 2015: 22)

The hylemorphic schema, then, is responsible for miring metaphysics in a substantialism that prevents an understanding of the becoming of being in which form is generated through an operation that is immanent to being itself. Such a processual understanding of ontogenesis is denied if form is understood as something that is imposed on an abstract matter from outside rather than involving inherent material processes.

Before developing Simondon's positive account of individuation it is first necessary to describe another technical schema, cybernetics, that also offered a solution to the problem of individuation. Cybernetics was of enormous importance for Simondon and his project involves the reworking of many of its core notions such as teleology, analogy, information, signification (meaning), vitalism and mechanism.

CYBERNETICS

Cybernetics is an interdisciplinary science whose core concern, at least in its first phase, was the study of the principles of organization and control in complex systems. As a caveat it cannot be stressed enough that to actually stipulate that there was a coherent first-order cybernetics programme in itself is problematic. The range of thinkers generally accepted as being first-order cyberneticians offered a range of perspectives that were not always harmonious.

Despite this heterogeneity there are some core concerns that all cyberneticians accepted (to varying degrees) or at least developed in new directions.

Rather than give a detailed account of a specific cybernetician's work I will attempt to present the core set of concepts that engaged all these thinkers in order to thereby illuminate Simondon's reimagining of this project. Although this approach will involve some historical material, the aim of this chapter is to locate Simondon's thought in relation to cybernetics as one of his major interlocutors.

The focus of cybernetic study is with the principles that functionally guide systems and how these principles can be applied to interpret other types of system. As such cyberneticians can be divided between those who hold that although cybernetics can account for purpose in machines it can only be applied to organisms *by analogy* and those who hold that there is an isomorphism (in some cases even an identity) between machine and organism to the extent that both can be explained via cybernetics.

For the first group cybernetics is not concerned with systems from a material perspective, that is, with what they consist of, but rather with their formal elements that can be used *analogically* to explain a broad range of systems. For the second group cybernetic explanation proposes that all systems literally operate cybernetically, that is, that both machine and organisms actually operate using the same principles.

Cybernetics is interdisciplinary in that its goal is to universally apply its core principles to explain systems hitherto subsumed under separate disciplines, such as biology, technology, psychology and sociology. The encyclopaedic scope of Simondon's work takes up this interdisciplinary approach, although utilizing a different axiomatic.

Of all the sciences, due to its affiliation with engineering (via Wiener) and mathematics (via Shannon), cybernetics is most closely allied to physics. However, cybernetics' goal is to be a general science that can account for phenomena described by all the other sciences using the new tools of information, modelling and control. As we will see, although also interested in engineering and information, Simondon's significant interest in biological science gives him some distance from much cybernetic theory.

TELEOLOGICAL MECHANISMS

To tease out several of the key concepts important for understanding cybernetics, and by extension Simondon, I will begin by looking at several of its foundational statements starting with Bigelow, Rosenblueth and Wiener's article 'Behavior, Purpose and Teleology', published in 1943. This article introduces many of the central concerns of cybernetics in embryonic form, namely teleology, technology and the behaviour of organisms.

In this article the authors reassess the importance of the notion of purpose by critiquing behaviourism for investigating its objects solely from the perspective of inputs and outputs and thus providing explanations so broad that behaviour comes to mean 'any change of an entity with respect to its surrounding' (Bigelow, Rosenblueth and Wiener, 1943: 18).

The authors then set about building a taxonomy of behaviours, beginning by distinguishing them as to whether they originate from input from outside the object (passive behaviour) or from within the object (active behaviour, where the object is the source of the output energy for the behaviour). The taxonomy then develops by discriminating between active behaviours as being either purposeful or purposeless.

> The term purposeful is meant to denote that the act or behavior may be interpreted as directed to the attainment of a goal—i.e., to a final condition in which the behaving object reaches a definite correlation in time or in space with respect to another object or event. (Bigelow, Rosenblueth and Wiener, 1943: 18)

Teleological, purposeful, active behaviour, being 'directed', requires the involvement of negative feedback, which is when the object's behaviour is controlled 'by the margin of error at which the object stands at a given time with reference to a relatively specific goal' (1943: 19). Behaviour is deemed to be purposeful if it is guided by signals from the goal at which it is directed. Or to put it in more cybernetic terminology, a system of auto-regulation between an entity and its goal is developed in which part of the output of the behaviour of the entity is returned to the entity as input.

This discussion of teleology is significant, not only because it introduces the concern with feedback, a core cybernetic concept, but also because the authors contextualize their position in relation to other notions of teleology and causality. An important reference point is Kant's concern with teleology and organisms in *Critique of Judgment*. A significant problem for Kant was how to resolve the antinomy between a universe governed by Newtonian causal mechanism and the seemingly indeterminate, yet purposive, behaviour exhibited by organisms.

The resolution Kant provides for this problem is the attempted reconciliation of Newtonian, efficient causality in phenomenal nature with the thinkability and actionability of final causality in the free behaviour of finite, rational beings.

The authors of this article demonstrate an awareness of this prior wrangling between final and efficient causality and claim to have avoided the pincers of Kant's antinomy by redefining teleology as an activity that is not subsumed under a final cause yet also evades strict subsumption to mechanical causality. This last point is crucial for it claims that teleological purpose is simultaneously congruent with efficient causality, yet also non-deterministic, as it is positioned in a new 'realm' both immanent to, and emergent from, the actual system in which it occurs:

> According to this limited definition, teleology is not opposed to determinism, but to non-teleology. Both teleological and non-teleological systems are deterministic when the behavior considered belongs to the realm where determinism applies. The concept of teleology shares only one thing with the concept of causality: a time axis. But causality implies a one-way, relatively irreversible functional relationship, whereas teleology is concerned with behavior, not with functional relationships. (Bigelow, Rosenblueth and Wiener, 1943: 22)

Here we can already discern how complexity and systems theory are nascent in cybernetics (even before it was called cybernetics) with the description of an emergent realm that escapes strict determinism and that is established via feedback mechanisms within systematic structures (this is further developed in second-order cybernetics regarding how the behavioural terminology of input and output get transformed into discussion regarding the inside and outside of systems).

Once the realm of teleology has been established as having explanatory force another significant aspect of cybernetic thinking is established: that it can be used to explain the behaviour of both machines and organisms. That is how it can answer the problem of natural purposes.

The authors make a brief comparison of the failed attempt of a human patient with a damaged cerebellum to drink a glass of water with that of the operation of a machine with an 'inadequately damped' feedback mechanism. From this they conclude:

> The analogy with the behavior of a machine with undamped feed-back is so vivid that we venture to suggest that the main function of the cerebellum is the control of the feed-back nervous mechanisms involved in purposeful motor activity. (Bigelow, Rosenblueth and Wiener, 1943: 19)

Through *analogy* is developed the possibility of an *isomorphism* between machine and organism in terms of teleology (understood as negative feedback). The extent of this isomorphism was by no means agreed upon among all the attendees of the Macy conferences,[2] but what is of concern to us here is that a physicalist account is being established to explain the behaviour of both machines and organism via purpose rather than behaviourism, dualism or finalism.

PHYSICALISM AND COMPLEXITY

While Wiener held that there was merely an analogous relationship between organisms and machines, that is, that the study of teleological mechanisms in one may be instructive when studying the other, Warren McCullough went beyond mere analogy and insisted on an identity between them. For him, organisms were just very complicated machines and should be described in the same manner.

In an article co-published with Walter Pitts (also in 1943) called 'A Logical Calculus of the Ideas Immanent in Nervous Activity', another aspect of the general cybernetic project received impetus that aimed to account for the physical mechanisms that were responsible for the mind.

Once again it must be stressed that there was some disagreement among the members of the cybernetic group regarding the place of mind in the description of organisms. Some of the group, such as Rosenblueth and Wiener, were eliminativist[3] in that they held that folk descriptions of mind shouldn't have a place in accounts of behaviour, which should be described in terms of control mechanisms and feedback.

McCullough and Pitts, however, had a more reductionist approach that described the neurons of the brain as embodying 'propositions' in a logical calculus. Broadly, McCullough and Pitts were defending the position that the brain was a logical calculating machine that operated using a binary neuronal mechanism. In their description the teleological behaviour that Wiener had described in his paper was situated at the level of the neuronal activity of the brain. Von Neumann (1966) introduced the notion of complexity into this account by acknowledging that the model developed by McCullough and Pitts could logically be used to explain a great deal of behaviour, but that complex systems also exhibited behaviour that could not be explained logically. However, that is not to say that these behaviours could not emerge from this system, just that their emergence would be too complex to be easily accounted for:

> Now this threshold of complexity, he supposed, is also the point at which the structure of an object becomes simpler than the description of its properties. In the usual case, which is that of simple machines, it is less complicated to describe in words what the automaton can do than to reproduce the structure of its wiring. For complex automata, the opposite is true: it would be simpler—indeed, infinitely more simple—to describe the structure of the automaton than to completely specify its behavior. (Dupuy, 2009: 141)

PURPOSE, RANDOMNESS AND INFORMATION THEORY

Another central notion of cybernetic theorizing that is of particular importance for Simondon is that of information. This concept became prominent with Claude Shannon's paper 'A Mathematical Theory of Communication' (Shannon, 1948), which was later popularized (along with the informational theories of Wiener and others) by Warren Weaver in his 'Recent Contributions to The Mathematical Theory of Communication' (Weaver, 1949). Although developed to describe the problem of communication from an engineering perspective, the theory also became prominent in certain areas of communication, media and cultural studies as well as having a significant impact in the biological sciences.[4] Shannon's theory of information is a development of Boltzmann's method of statistically calculating entropy in a thermodynamic system, applied to the context of communication. For Shannon, entropy in this context becomes the measure of uncertainty surrounding the communication of a message. As such, this theory is mathematical and is not concerned with whether what is being communicated has any semantic content. As Weaver points out:

> In fact, two messages, one of which is heavily loaded with meaning and the other which is pure nonsense, can be exactly equivalent, from the present viewpoint, as regards information[,] (Weaver, 1949)

the 'present viewpoint' being that at which communication is described by statistical probability.

There are two aspects to Shannon's theory that are worth analysing here: the first is the definition of information as statistical probability and the second is the concept of noise. Shannon's initial work on communication was in cryptography and it is easy to see how this influences his theory of information in that what it describes is the probability of one string of symbols in a message occurring instead of another. The measure of information is calculated by realizing that all messages occur via selection from a set. As a set is finite, the probability of any message occurring, especially when taking into account any previous communication, is thus calculable.

This calculating operation was likened to the working of a transducer: 'a device capable of decoding and recoding strings of symbols as inputs and outputs, one that "may have an internal memory so that its output depends not only on the present input symbol but also on its past history"' (Shannon and Weaver, cited in Mirowski, 2002: 71).

It is not difficult to see how such an operation could be useful both for mathematical operations in cryptography as well as for describing a way to measure information. For such a statistical operation to occur, as well as communication, there must be a system in place that consists of a sender and receiver that both use the same set of possible messages. As Ashby writes:

> Communication thus necessarily demands a set of messages. Not only is this so, but the information carried by a particular message depends on the set it comes from. The information conveyed is not an intrinsic property of the individual message. (Ashby, 1957: 124)

Information is therefore a relation between a message and redundancy (the range of unselected messages in the set or code). In Shannon's mathematical theory this relationship is strictly non-semantic and merely refers to the possibility of there being content in a communication. From an engineering standpoint this statistical measure of information is extremely useful in overcoming the problem of noise; that is, it solved the problem of how a message from a sender may accurately be sent to a receiver without any distortion introduced by the channel along which the message passes.

As well as enabling a message to be distinguished from noise, the theory of information also enables the ability to disguise a message within noise as occurs in cryptography. These are, as it were, two sides of the same coin and involve information as a statistical measure of probability within a code.

Although Shannon's focus was on applying his theory of information to problems of communication in engineering (Terranova, 2004: 29), the significant leap for cybernetics was that taken by Wiener in considering information as negative entropy in other domains outside of engineering such as biology and social systems (Wiener, 1989).

Ashby explains the importance of this leap through the example of the role of information in the growth of a rabbit ovum. From the informational perspective the analysis of the growth of an ovum using energetics does nothing to answer the question of why this ovum develops into a rabbit rather than some other form. In the same way that information theory measures the probability of the selection of a single message from the set of all possible messages,

> Cybernetics envisages a set of possibilities much wider than the actual, and then asks why the particular case should conform to its usual particular restriction. . . . So no information or signal or determining factor may pass from part to part without its being recorded as a significant event. Cybernetics might, in fact, be defined as the study of systems that are open to energy but closed to information and control—systems that are 'information-tight'. (Ashby, 1957: 3)

In this case the selection relates to which form the ovum will take as it develops. This selection is a function of information and, as such, is mathematically calculable, just as a message is calculable given its code using information theory. Thus the thermodynamic notion of entropy informing a definition of information is then fed back to provide an explanation of negentropy in physical systems. The notion of information has thus migrated from being a measure of statistical probability in the transmission of messages to a causal input within a system, and as such must also have a material reality. As Terranova notes:

> This notion that information was somehow related to anti-entropic or negentropic forces is at the basis of the informationalist perspective that identifies information with a kind of form determining the material unfolding. (Terranova, 2004: 31)

For cybernetics, therefore, it is evident that there is a close connection between information and the production of form (including behaviour) via the link with entropy. As Mirowski observes, there is a relationship here between the idea of Maxwell's famous demon as theorized in thermodynamics, Wiener's anti-aircraft weapon and the notion of negentropy in nature:

> [A] demon taught to neutralize the devious codes of the Enemy now trains his strategic knowledge on Nature to defeat the forces of dissolution and disorder. (Mirowski, 2002: 73)

Where Wiener began his study of teleological mechanisms by developing a machine to track and destroy enemy aircraft through increasing the accuracy of prediction, he ends by asserting that it is this same mechanism of information that allows organisms to not only produce their own teleology but to actually develop in the face of the second law of thermodynamics in the first place.

ORGANISM AND MACHINE

It is perhaps unsurprising that cybernetics developed a longstanding fascination with automata and machines given that Wiener developed his teleological theory from the operation of a machine (the anti-aircraft gun).

The types of machines that most interested the cyberneticians were those whose operation involved some form of self-regulation via feedback loops. Some of the more famous examples are Claude Shannon's maze-negotiating mouse, which could actually remember how to navigate a particular maze once it had solved it; Ross Ashby's Homeostat, which was a machine that purportedly demonstrated how a system could maintain homeostasis within a changing environment; and, of course, Von Neumann's work on natural automata, which led to the development of the modern computer.

It is the formal operations or behaviours of systems that interested cybernetics and as we have seen these rely not on mechanistic determinism but on feedback mechanisms. That is, cybernetics is concerned with 'forms of behaviour in so far as they are regular, or determinate, or reproducible' (Ashby, 1957: 1). That this theorizing was extended by some to explain the behaviour of organisms is also unsurprising given the material independence information was understood to have by many, as Ross Ashby states:

> Cybernetics, too, is a 'theory of machines', but it treats, not things but ways of behaving. It does not ask 'what is this thing?' but 'what does it do?' (Ashby, 1957: 1)

One outcome of this ecumenism was the development of modelling behaviour (comprehended using the theories of feedback and information) across and between what had been seen as separate domains of knowledge. As Dupuy concisely explains,

> To abstract the formal properties of phenomena and in this way identify isomorphisms between different domains of phenomena is precisely what modelling is all about—even science itself. The attempt to propose a unified theory of machines and living creatures with reference to the category of purpose, conceived in mechanistic terms and rebaptised as teleology represented a spectacular increase in the extension of science, hardly a rupture with it. (Dupuy, 2009: 47)

VITALISM AND MECHANISM

Wiener argued that it is this extension of science that enabled a new way of understanding the operation of purpose in organisms that avoided both vitalism and Newtonian mechanism. The overcoming of this dualism was achieved by the development of new types of complex mechanisms that were able to act without recourse to folk psychology and consciousness. As Wiener triumphantly proclaims,

> Vitalism has won to the extent that even mechanisms correspond to the time-structure of vitalism; but as we have said, this victory is a complete defeat, for from every point of view which has the slightest relation to morality or religion, the new mechanics is fully as mechanistic as the old. (Wiener, 1965: 44)

There are two attitudes that cyberneticians tend to have regarding consciousness and psychology, both of which stress the transcendental priority of the material to mental phenomena. The first attitude is reductionist (like McCullough's), in which mental phenomena are viewed as inconsequential epiphenomena produced by complex material mechanisms. The second allows significance to consciousness but only because it is a consequence of mechanism and thus it is the mechanism that is of fundamental interest (there is a sense that structuralism emerges from this line of thinking). Both attitudes share the same basis, however, which is that what is primary is mechanistic, thus consciousness and meaning are fundamentally underpinned by mechanisms of control and communication.

This perspective can still be aligned with Kantianism to the extent that what has been added is only the contention that there is a mechanistic explanation for the emergence of the structuring of experience and understanding.

It is in this sense that Dupuy (2009) argues that what cybernetics, and cognitive science after it, are doing is not so much trying to emulate the mind using machines but to make the much bolder move of claiming that the mind *is* a machine. Such an identity is clearly evident, for example, in W. Ross Ashby's *Design for a Brain*, in which he attempts 'to deduce from the observed facts what sort of a mechanism it [the brain] must be that behaves so differently from any machine made so far' (Ashby, 1954: v). The opening of the third chapter called 'The Animal as Machine' makes explicit this assumed identity:

> We shall assume at once that the living organism in its nature and processes is not essentially different from other matter. The truth of the assumption will not be discussed. (Ashby, 1954: 29)

As such we can see that the history of cybernetics undertook a shift from merely comparing animals with automata analogically to making the much stronger claim that animals *are* machines. A whole series of activities emerged from this mechanizing cybernetic movement, including second-order cybernetics, with its subsequent development into systems theory and neurophenomenology; artificial intelligence, which moved beyond behaviourism to theorize mental states and the nature of how a mind could be instantiated machinically; microbiology, which theorized genes as informational units; and cognitive science, which viewed cognition as a form of information processing.

In conclusion to this section, and with some generalization, we have seen that cybernetics claims purposeful behaviour and finality in organisms and machines can be explained in terms of physical laws. However, these physical laws are not those of the deterministic kind found in Newtonianism, but consist of the operations of feedback augmented by information. Additionally, the quantitative nature of information means all behaviour can potentially be modelled mathematically and organic behaviour should be able to be reproduced mechanically, as it is the form of behaviour that is important and not the material in which that behaviour occurs. This final point is an important one as it enables cyberneticians to claim identity between machine and organism, something that Simondon strongly disputes.

SIMONDON AND CYBERNETICS

Undoubtedly Simondon was heavily influenced by cybernetics but there are several important aspects of his work that distinguish it from the main currents of the early cybernetic project described above.

That Simondon was involved with the cybernetic movement is evidenced by the fact that in 1964 he organized a conference in Royamount, entitled *The Concept of Information in Contemporary Science* (1965), with Wiener as an invited speaker. As John Hart notes in his preface to the 1980 translation of *On the Mode of Existence of Technical Objects*, at this meeting Simondon compared cybernetics to the work of Newton due to its universal ambition:

> In fact, historically, cybernetics appeared as something new directed to achieving a synthesis; in sum, we find ourselves brought back to the time of Newton, or to the time when the great philosophers were mathematicians or scientists in the natural sciences and inversely. This is doubtless the context in which it is now possible to listen to what Professor Wiener has to present to us. (Simondon quoted in Hart, 1980)

Simondon is not critical of the synthesizing ambition of the cybernetic project; in fact his own axiomatizing project (which he calls *allagmatics* [Simondon, 2013: 529]) has a similar encyclopaedic ambition, which is to be achieved via a revision of cybernetic theory. In the following section I will look at some of Simondon's criticisms of cybernetics in order that we can more clearly describe his own project of reformation.

THE CRITIQUE OF HYLEMORPHISM AND ATOMISM AS APPLIED TO CYBERNETICS

Where cybernetics is predominantly concerned with behaviour, Simondon focuses on operation. The operation on which he focuses is always one that prioritizes the process and nature of individuating relation and how terms come to be constituted. Indeed, it is cybernetics' reliance on describing mechanisms from the perspective of already constituted individuals that is the basis for one of Simondon's criticisms. For him, that a message always travels between a sender and a receiver begs the question of how these entities came to be established in the first place. It is this question, essentially one of ontogenesis, that drives Simondon's entire philosophy.

For Simondon, an account of ontogenesis (the process of becoming of entities described as individuals) is fundamental. By denying it we are eliding the key issue of what an individual is and how it comes into existence.

The criticisms of hylemorphism and atomism discussed earlier are both applied to cybernetics because it focuses on already individuated systems without giving an adequate account of their genesis. For this reason Simondon disagrees with any prioritization of the importance of information *as message* within an already individuated system. This is not to say that he contests information as it is theorized by cybernetics, only that it is of primary importance. In a footnote he makes it clear that he sees cybernetic information as very much secondary and reliant on a more fundamental and primary operation which he calls 'primary information':

> This affirmation does not lead us to contest the validity of the quantitative theories of information and the measurements of complexity, but it supposes a fundamental state—that of the pre-individual being—prior to any duality of the sender and of the receiver, and therefore to any transmitted message. It is not the source of information that remains of this fundamental state in the classic case of information transmitted as a message, but the primordial condition without which there is no effect of information, and therefore no information: the metastability of the receiver, whether it be technical being or a living individual. We can call this information 'primary information.' (Simondon, 2010b: 15)

It is this 'primary information' that Simondon describes as transduction and that he offers as the foundation of his theory of ontogenesis, which I will examine in the next chapter. For the purpose of this chapter it is enough to note that the cybernetic understanding of information is susceptible to Simondon's critique of hylemorphism.

Although Simondon's notions of individuation and information share the same thermodynamic heritage as the cybernetic notion of information, they are defined in a very different way. As we have seen, cybernetics engages with thermodynamics by reconfiguring the mathematical measure of entropy into the negentropic measure of information. Simondon's engagement with thermodynamics, however, involves the development of the role of thermodynamic metastability but moves away from the cybernetic concern with quantifiable information.

Just as Simondon takes issue with the account of pre-constituted individuals between which information travels, so he also questions the quantitative theory of information as adapted from Shannon. For Simondon this account of information falls foul of the criticisms he lodged against both atomism and hylemorphism. It is atomistic whenever it is described as discrete quanta transmitted between relata with no account of how such information comes to be individuated. Additionally it also leaves unanswered how the terms between which such information is transmitted also come to be individuated. The cybernetic conception of information can also be understood hylemorphically as a source of form separate from the matter being formed, rather than form arising from the individuation process itself.

As we will see in the next chapter, for Simondon information is not to be considered as a message between a source and receiver but describes an operation relating to the resolution of a disparity within being leading to the individuation of structure.

ANALOGY AND ALLAGMATICS

Analogy holds an important place for Simondon, as it did for the cyberneticians, in the development of an epistemology. Due to his refusal to prioritize already constituted individuals, Simondon bases his epistemological method on the analogy of operations, a method he calls *allagmatics*.

Simondon develops his epistemology by first making a clear distinction between structure and operation and the clear difference regarding the kind of knowledge that we can derive from studying either. His contention is that scientific and traditional analytical epistemologies are too concerned with already individuated structures, which they tend to understand as static and substantial, while overlooking the constantly individuating, processual and operational nature of reality. As such what is required is a redressing of the balance:

> The duty of the allagmatic epistemology is to determine the true relationship between the structure and operation in being, and thus to organize the rigorous and valuable relationship between structural knowledge and the knowledge of the operation of being, between analytical science and analogical science. (Simondon, 2013: 535)

It is the operation of individuation, Simondon argues, that is prior to any structuration and thus any focus on structure without reference to this operation is unbalanced, hence the need for an analogical science to complement analytical science. By analytical science Simondon means the reductionist method that 'assumes that everything is reducible to the sum of its parts or the combination of its elements'. Contrary to this, analogical science supposes that 'the whole is primordial and is expressed by its operation, which is a holistic function' (Simondon, 2013: 535). With the reference to holistic functioning Simondon is making an explicit commitment to the existence of a cybernetically inspired form of teleological causation.

However, although Simondon shares the cybernetic fascination with analogy, he is critical of the way that it implements it. His critique is important in that through it we can understand what Simondon considers a valid use of analogy.

> Analogical thought is that which observes identities of relations, not relations of identity but it must clarify that these identities of relation are the identities of operative relations, not the identities of structural relationships. By itself it discovers the opposition between resemblance and analogy: resemblance is given from structural relationships. Pseudo-scientific thought makes substantial use of resemblance, sometimes even the resemblance of vocabulary, but it does not make use of analogy. (Simondon, 2013: 533)

For Simondon analogy is only valid 'if it covers a world where beings are defined by their operations and not by their structures' (Simondon, 2013: 534), for an analogy based on structures does not in fact constitute an analogy for Simondon but merely a resemblance because it 'can't reach the whole reality of being'. Simondon's theory of knowledge is thus premised on his ontological account of the operation of being. Thus for Simondon, Kant's theory of knowledge, as it focuses on the structure rather than operation, is based on resemblance and is therefore limited.

It is perhaps worth reiterating that cybernetics' prime focus is in looking at form, particularly 'forms of behaviour in so far as they are regular, or determinate, or reproducible' (Ashby, 1957: 1). Cyberneticians are interested in drawing analogies between phenomena that can be interpreted as enacting the same kind of behaviour though involving different types of system. For example, analogies might be drawn between the working of the brain and a computer, or cellular automata and living organisms, or between games and the economy, as well as to applying the findings from one system to that of the other. It is in this spirit that analogies were often attempted between the behaviour of robotic automata and that of living beings.

It is clear then that, for Simondon, such analogies of *form* amount to mere structural resemblance and are not valid analogies relating to operation. Although a cerebellum may cause the same kind of behaviour in an animal as a servo-mechanism does in an automaton, this is just a resemblance and tells us little about their differing *operative* realities.

In the text 'Technical Mentality', Simondon (2009) discusses how, through the use of analogy and paradigm, technology has aided the development of different 'modes of knowledge'. We have already seen that the schema of hylemorphism was developed from the process of moulding. Here he gives two further examples of the development of such universal 'schemas of intelligibility'. In the first, the operation of simple mechanisms offers an analogy for how the rigorous logic of Cartesian mechanism operates (Simondon, 2009: 17). The implication is that the schema for the operation of thought, that is, how the understanding operates, has developed analogically with the operation of machines. Thus in Cartesianism both machines and thought are understood rationally and with an operation consisting of a 'transfer without losses', which is to say substantially. Thus,

> the only domains that are accessible to philosophical reflection are those with a continuous structure. It will therefore be clear why one has wanted to consider living beings as machines: if they weren't machines ontologically, they would have to be so at least analogically in order to be objects of science. (Simondon, 2009: 18)

This is precisely the problem that troubled Kant in his third critique in attempting to understand natural purposes using Newtonian mechanism.

Cybernetics is the second example Simondon gives of a universal schema for the operation of the understanding developed from an understanding of technical operation. The key concept here is the regulatory feedback mechanism 'that allows for an active adaptation to a spontaneous finality' (Simondon, 2009: 18), which Wiener understands as teleological mechanism. As Simondon observes, this schema has proven useful when applied analogically to other phenomena:

> This technical realization of a finalized conduct has served as a model of intelligibility for the study of a large number of regulations—or of regulation failures—in the living, both human and nonhuman, and of phenomena subject to becoming, such as the species equilibrium between predators and preys, or of geographical and meteorological phenomena: variations of the level of lakes, climatic regimes. (Simondon, 2009: 18)

In a revealing paragraph Simondon further explicates the problem that analogy presents for a technical schema such as cybernetics when applied universally:

> In this sense, technology manifests in successive waves a power of analogical interpretation that is *sui generis*. . . . None of the schemas exhausts a domain, but each of them accounts for a certain number of effects in each domain, and allows for the passage of one domain to another. This transcategorical knowledge, which supposes a theory of knowledge that would be the close kin of a truly realist idealism, is particularly fit to grasp the universality of a mode of activity, of a regime of operation; it leaves aside the problem of the atemporal nature of beings and of the modes of the real; it applies to their functionings; it tends towards a phenomenology of regimes of activity, without an ontological presupposition that is relative to the nature of that which enters into activity. Each of the schemas applies only to certain regimes of each region, but it can in principle apply to any regime of any region. (Simondon, 2009: 18)

Thus although the schemas of Cartesianism and cybernetics are valid in some domains and can, in principle, be analogically applied universally, this application will be invalid in some instances as they, to some extent, neglect the ontological nature of that to which they are applied. The move that Simondon makes is to insist on an 'ontological presupposition' by which analogies can be more precisely grounded in operation; it will be such an ontological presupposition that Simondon brings in to play as he seeks to both underpin and extend his own account of a schema of intelligibility.[5]

A pertinent concept Simondon uses in the above quote is that of a 'realist idealism'. In this context it refers to the epistemological use of technological operations for providing the operational material for schemas of knowledge. With hylemorphism, Cartesianism and cybernetics, as we've seen, the analogies are often poor. However, what Simondon is proposing is that such a technical mentality, although currently insufficient, can be developed further and more precisely.

It is with Simondon's development of the transductive operation in which the epistemological force of his project is felt. Not only is it the case that Simondon develops an axiomatic operation pertinent for the individuation of being, but that it is also applicable to the individuation of thought given thought is a phase of being. Thus Simondon's idealism is realist because to know is itself an operation of individuation, that is, of thought tracking the individuation of being. Thus he writes:

> We cannot, in the usual sense of the term, know individuation; we can only individuate, individuate ourselves, and individuate in ourselves; this seizure is therefore in the margins of knowledge strictly speaking, an analogy between the two operations, which is a certain mode of communication. (Simondon, 2013, 36)

This striking move of tracking the individuation of thought as it itself tracks the individuation of being has a certain resonance with the later second-order cybernetics. With this method Simondon will navigate a path between empiricism and Kantian idealism. The rigid categories of Kant are replaced by dynamic schemas that individuate from the 'phenomenologies of regimes of activity'. As such one can identify a kind of Platonism in Simondon in which technical schemas represent modes of operation that are, in a sense, eternal in that they can be repeatedly revived and instantiated in technical structures. They are also episodes in the progressive individuation of a history of ideas, which will themselves be instantiated in, as well as productive of, further invention.

There is the intertwining of a combination of feedback loops at the heart of this epistemology, including the operation of the individuation of thought alongside the individuation of being in order to understand the nature of that individuation. Such an operation of thought is also, at the same time, an individuation for thought in that it develops the schemas thought has at its disposal to understand the world. Additionally, this individuation can also be a subject for reflection and an evaluation of the appropriate application of these schemas.

> Individuation of the real exterior to the subject is seized by the subject thanks to the analog individuation of knowledge in the subject; but it is through the individuation of knowledge and not only by the knowledge that the individuation of beings not subjects is seized. (Simondon, 2013: 36)

A simple example of how the tracking of individuation by thought rather than the application of knowledge onto beings can help identify a poor analogy is given in *On the Mode of Existence of Technical Objects*, in which Simondon writes:

> There is one element that threatens to make the work of Cybernetics to some degree useless as an interscientific study (though this is what Norbert Wiener defines as the goal of his research), the basic postulate that living beings and self-regulated technical objects are identical. The most that can be said about technical objects is that they tend towards concretization, whereas natural objects, as living beings, are concrete right from the beginning. There should be no confusing of a tendency towards concretization with a status of absolutely concrete existence. (Simondon, 1980: 41)

Once again Simondon applies the argument regarding the genesis of the individual. By stipulating that natural, living beings are 'concrete right from the beginning' he is pointing to a fundamental difference between the ontogenesis of living and technical individuals. A technical individual might be more concrete than a previous member of the same technical lineage but it could not achieve this development by itself, whereas 'a living being engenders other beings that are similar to itself or that can become like it after a certain number of successive reorganizations that occur spontaneously if the conditions are suitable' (Simondon, 1980: 61).

The full force of this argument only becomes apparent when we more fully understand Simondon's operational ontology and how it leads to a very different understanding regarding the individuation of biological organisms and the development of technical lineages via progressive concretizations. Cybernetics' obsession with automata as a special class of technical objects imbued with a particular analogical power also strikes Simondon as mistaken in that the bad analogy not only leads to a mistaken conception of the organism, but also to an impoverished understanding of what technology is:

> It would not even be right to found a separate science for the study of regulatory and control mechanisms in automata built to be automata: technology ought to take as its subject the universality of technical objects. In this respect, the science of Cybernetics is found wanting; even though it has the boundless merit of being the first inductive study of technical objects and of being a

> study of the middle ground between the specialized sciences, it has particularized its field of investigation to too great an extent, for it is part of the study of a certain number of technical objects. Cybernetics at its starting point accepted a classification of technical objects that operates in terms of criteria of genus and species: the science of technology must not do so. There is no species of automata: there are simply technical objects; these possess a functional organization, and in them different degrees of automatism are realized. (Simondon, 1980: 49)

That cybernetics focused on the narrow range of technical objects that it understood as utilizing feedback and information for regulation indicated to Simondon the narrowness of its concerns despite its attempted universality. That cybernetics also failed to be interested in the mode of individuation of this restricted class of technical objects, even while contrasting them with organisms, compounded its mistake. Two specific areas that Simondon saw as troubling for this attempted analogy, as well as the associated cybernetic notion of information, were those of meaning and memory.

MEMORY AND MEANING

Simondon argues that the operation of memory in humans and machines is fundamentally different. Where machine memory 'is able to retain very complex, detailed, precise monomorphic documents for a very long time' (Simondon, 2012: 75), human memory involves the retention of forms and meaning that are significant for that individual. What's more, human memory is plastic in that new content introduced to memory will change how it understands future content: 'in the living the *a posteriori* become *a priori*; memory is the function by which *a posteriori* matters become *a priori*' (Simondon, 2012: 77).

Simondon thus understands epistemology as a function of ontology insofar as the development of schemas of ontological operation (such as those developed from technology) enables developments in epistemology. Such a theory is in stark contrast to the rigid categories of Kantian transcendental idealism for it allows for the development of progressive schemas of intelligibility whose understanding may well cut across the rigidly defined Kantian categories.

This difference between the operation of machine and human memory also leads to the impossibility for analogy between the operations of self-regulation in each:

> Machine memory is that of the document, of the result of measurement. Human memory is that which, after an interval of many years, recalls a situation because it involves the same meanings, the same sentiments, the same dangers as another, or simply because this similarity has a meaning according to the

> implicit vital coding that constitutes experience. In both cases memory allows self-regulation; but human memory makes self-regulation possible according to a set of meanings that are valid in the living and that can develop only in the living; machine memory establishes a self-regulation that has meaning in the world of non-living beings. (Simondon, 2012: 78)

Such a distinction puts into question the account of teleology (negative feedback) described by Wiener as being a suitable account for the kind of self-organizing activity that Kant describes for natural purposes (organisms). This difference in the operation of memory and of self-regulation highlights another significant problem for cybernetics and that is regarding semantics.

First, in the cybernetic account of teleology the origin of meaning remains unclear. For example, although the notion of the goal of a system is used to describe the meaning of an act, this can be seen to be a sleight of hand in that it is the observer or creator who is allocating a sense of meaning to the system (by analogy) rather than meaning emerging immanently from the system itself. In short, meaning would either be epiphenomenal ghosting or requires a physical explanation just as any other phenomenon does.

Second, as already mentioned, the lack of an account of semantics at the heart of the Shannon-Weaver account of information as probability makes it problematic as an explanation of communication. Mirowski discusses this lack:

> In Shannon's version, there is no macro-micro distinction, only a given probability of a particular symbol showing up, and a measure of the likelihood of strings of symbols. This is often rephrased by suggesting that Shannon's entropy is about 'choice' of symbols (Hayles, 1990a: 54), an interpretation first popularized by Weaver: 'information, in communication theory, is about the amount of freedom of choice we have in constructing messages' (Shannon and Weaver, 1949: 13). (Mirowski, 2002: 72)

Mirowski goes on to add that this idea of choice is 'sneaking intention and semantics in through the back door' (Mirowski, 2002: 72). After all, who is doing the choosing?[6]

A significant aspect of this inability for the cybernetic notion of information to develop meaning is that it fails to describe a system as being in a true relationship with what is external to it. As we will see it is the openness of a system onto a milieu that is required for Simondon to enable a true theory of invention and emergence. It is also this openness that is required for the development of any system whether it be a language or a higher organism to develop meaning.

This lack of a coherent account of meaning in first-order cybernetics also became an issue for the projects of AI and the Philosophy of Mind that followed. As Philippe Breton writes, 'Cybernetics has been one of the principal destabilizing instruments of the anthropomorphic conception of man. . . . Cybernetics therefore assumes [the form of] a terrible paradox: it affirms humanity while at the same time depriving man of it' (Breton quoted in Dupuy, 2009: 109).

More specifically, the paradox cybernetics poses is that the more knowledge it claims to have of the physical and causal mechanisms from which the human is constructed the further the place of the human subject and its own universe of meaning is undermined as being a mere fiction, an illusion that hides those true mechanisms at work beneath.[7]

CONCLUSION

At the beginning of this chapter I mentioned that Simondon's work had an encyclopaedic scope, something that it also shared with cybernetics. As we've seen, the aim of cybernetics, broadly speaking, was the explanation of phenomena in different domains through information and feedback. Simondon's goal was to develop a science of operations (allagmatics), initially by categorizing them so as to 'define the major categories of operations, the different types of dynamic transformations that objective study reveals' (Simondon, 2013: 529), with the ultimate goal of defining 'a single fundamental type of operation . . . of which every particular operation is derived like simpler cases' (Simondon, 2013: 529). This latter goal, the search for the single axiomatic type of operation, will result in his reformulation of the notion of information as operation. His debt to cybernetics is clear when he states:

> Allagmatic theory is the study of individual being. It organizes and defines the relation of the theory of operations (applied cybernetics) and of the theory of structures (deterministic analytical science). (Simondon, 2013: 535)

However, the use of the term *cybernetics* here is indicative of his having transformed it so that information no longer refers to the passage of messages between entities or the measure of entropy but to the operation of individuation as an ontology.

This operational ontology also includes the production of a new analogical epistemology. One of its goals is to map the types of operation that occur in being (physical, biological, psychosocial, technical). It should perhaps be noted here that Simondon is not attempting a refutation of an account of analytical science (for example, by portraying it as reductionist) but seeking to point out an operative aspect of reality that science overlooks.

It is arguably the case that since Simondon wrote these words (he wrote his thesis in the mid-1960s), this suggested project of mapping types of operation has indeed become part of scientific investigation. The science of complexity has certainly made progress with describing processes in physical phenomena, as we will see in the next chapter.

Although Simondon describes allagmatics as being 'in the order of the sciences', he certainly does not mean that this restricts it to what are seen as the classical sciences (physics, chemistry, biology). As will be investigated in the coming chapters, allagmatics is just as relevant to investigations of the social, technical and aesthetic. It is as such that Simondon considers his project as encyclopaedic.

With the mapping of types of operation Simondon's can also be seen as a precursor to the kind of project that Manuel DeLanda describes when he sees the task of the modern philosopher as detailing the ontological reality of possibility spaces that are real even if not actualized:

> What is needed is a way of specifying the *structure of the space of possibilities* that is defined by an entity's tendencies and capacities. A philosopher's ontological commitment should be to the objective existence of this structure and not to the possibilities themselves since the latter exist only when entertained by a mind.[8] (DeLanda, 2011: 5)

The proposed mapping of types of operation is an aspect of the project of realist idealism that is rooted in Simondon's reformed notion of information, to which I now turn.

NOTES

1. Gestaltism is another theoretical explanation for the attainment of form that Simondon finds lacking. This will be addressed in chapter 3 in the discussion of perception given that it is essentially a theory of meaning.

2. The Macy conferences were a series of ten conferences that took place between 1946 and 1953 and were so called because they were sponsored by the Josiah Macy Junior Foundation. The participants came from a wide number of disciplines including computer science, psychiatry, anthropology, mathematics and sociology. The aim of the conferences was to found a general science of the human mind and cybernetics was the result. According to Dupuy (2009: 120), Simondon read the transactions of these conferences.

3. Eliminativism is the reductionist view that a more precise account of the mind can be given through scientific description of its physical basis (for example, at the neuronal level, via feedback mechanisms) than via folk descriptions (such as beliefs and desires). In its more extreme form, such as that proposed by the Churchland's, folk descriptions are viewed as simply mistaken and having no foundation.

4. Although there is some evidence that Shannon was interested in applying his theory to living organisms, Wiener maintained that this was his idea (see Mirowski, 2002: 70).

5. In the same essay Simondon describes this schema as being based on utilizing the notions of thresholds and networks. I will develop Simondon's account of these notions and his philosophy of technology over the coming chapters.

6. Merleau-Ponty also points out the problem of semantics for cybernetics when contrasting the notion of language as code in cybernetics to that of orthography found in a spoken and written language: 'Now cyberneticists never study the relation of the signifier and signified. The problem for them is to translate. We code the message and this is the fundamental operation. We still have to arrive at the moment when the message has a relation with what it means.

The enumeration of possible combinations does nothing to help us understand the very act by which language takes on a meaning' (Merleau-Ponty, 2003: 163). This leads him to the same conclusion found in Simondon: 'The code is no more a language than is the automaton a life' (Merleau-Ponty, 2003: 163).

7. The Churchland's project of eliminativist materialism is a good example of an attempt at this reductionist approach.

8. DeLanda's project uses concepts from the science of complexity (attractor, singularity, gradient, etc.) to map the various patterns by which various pre-individual conditions can develop. These patterns are described as spaces-of-possibility and are real (otherwise how could they be mapped?) although not actual. As such they have a similarity to Simondon's description of technical schemas as constituting a *realist idealism*.

Chapter Two

Physical Individuation

Transduction and Information

Having explored the more critical aspect of Simondon's philosophy, in this chapter I will turn to his positive ontogenetic theory of individuation. This theory is the heart of Simondon's project and requires the elucidation of a number of concepts that together underpin a unique understanding of ontogenesis; these include the key notions of the pre-individual, transduction, milieu and information. Additionally, I will also compare this account to some contemporary work in the study of self-organization, both in order to illuminate Simondon's theory as well as to help evaluate it as an axiomatic distillation of the processes that are associated with some recent work in complexity.

THE PRE-INDIVIDUAL

Before expanding on the operative nature of individuation it is first necessary to elucidate Simondon's conception of being (a less loaded term would be *primary reality*), for this is foundational for what follows. As previously established, Simondon argues that being must not be equated with either a fundamental substance (or substances) or a receptive matter that can be given form through the action of forces. Rather, he argues that 'primary reality is pre-individual' (Simondon, 1964: 127). It is important to be clear that Simondon is not here replacing the term *substance* with a synonym. The pre-individual does not name any primary entity or substance as such, but rather a *condition* of being, that is 'a being that is more than unity and more than identity, and that has not yet dephased itself into multiple dimensions' (Simondon, 2010b: 23).

It is clear that if we're not to fall into a retrogressive dualism (in which each term of the dualism requires further elaboration), a monism that provides a description of the dynamic production of entities in the world must account for that dynamism immanently. That is to say that being must always already be in a condition of falling out of phase with itself, that is, that its unity carries within itself the going beyond of itself.

> The sole veritable monism is that in which the unity is seized at the time where the possibility of a diversity of operation and structures is approached. The sole veritable monism is that in which, rather than following an implicit dualism that it seems to refuse, contains within itself the dimension of a possible dualism, but against a ground of being that can not slip away. . . . Dualism can only be prevented if one starts from a phase of being anterior to individuation, to relatavize individuation by situating it among the phases of being. (Simondon, 2013: 266)

The inspiration for this 'hypothesis of a state of pre-individual being' (Simondon, 2013: 317) comes predominantly from physics[1] and in particular the thermodynamic notion of metastability that describes a state that is neither completely stable nor unstable but somewhere between, that contains enough potentiality that it can 'produce a sudden alteration leading to a new equally metastable structure' (Simondon, 2013: 317).

An example of metastability that Simondon provides is the wave-particle duality of a photon, which can be simultaneously described as 'in a certain sense' a physical individual and also as an amount of energy that is open to transformation (Simondon, 2013: 317). Although physics does not give us the pre-individual in this example, Simondon contends that it 'does show that there are geneses of individualized realities starting from state conditions'. By necessity this state cannot be wholly stable; for it to be so would mean that it were fully substantial (in the Aristotelian sense). Instead it contains within itself the potential for transformation and hence it is always 'more than unity and more than identity'. The photon is then always more than just a particle and more than just a wave.

It is for this reason that Simondon's concern is not to ask what being *is*, such as seeking to uncover physical entities that can be said to constitute primary reality (such as quarks or Higgs-Boson) but to inquire into how primary reality *must be*. And for being to be able to change or become it must necessarily have the potential to be productive of disparity, which leads to the assertion that primordial being must be metastable.

For Simondon potential does not mean the same as possibility or the virtual but something wholly real that is indicative of the potential energy inherent in metastability: 'potential, conceived as potential energy, is the real, because it expresses the reality of a metastable state, and its energetic situation' (Simondon, 2013: 68). However, this should not lead us to equate the pre-individual with energy. That is, the pre-individual and energy are not equivalent, a fact that should be obvious when one considers that energy = mass x velocity² and therefore presumes prior individuation. So although energy figures significantly in Simondon's ontology it is not prior to form. Indeed energy and matter are two complementary dimensions that arise from the individuation of the pre-individual; they are 'manifestations' of the pre-individual but we are unable 'to seize the pre-individual real which subtends this transformation'.

Although Simondon's inspiration for the notion of the pre-individual is predominantly scientific, it is described in distinctly metaphysical terms, at times likened to the apeiron described by Anaximander.[2] Therefore although Simondon's metaphysics draws heavily on physical and biological science, the ground remains distinctly philosophical and metaphysical in character and not susceptible to scientific reduction. Hence the pre-individual is not a scientific concept but a philosophical one.

> Through its manifestations, that is to say when it changes, we perceive only the extreme complementary aspects; but these are dimensions of the real rather than the reality that we perceive; we grasp its chronology and its topology of individuation without being able to seize the pre-individual real which subtends this transformation. (Simondon, 1964: 130)

One of the roles such a notion enables is the thinking of an ontogenetic process that does not start from the assumption of already constituted individuals, which was, as we saw, one of Simondon's main criticisms of the theory of hylemorphism. What has been identified is a primordial condition from which all ontogenetic operations begin and which is maintained in aspects of all individuated entities thus enabling further operations of individuation.

Therefore we don't find in Simondon a dualism between becoming and being. Rather, as we saw in his proposed allagmatics, he makes a distinction between operation and structure, but is clear that operation is prior to the attainment of structure and what requires explanation are the operations involved when one structure is transformed to another. To better grasp the modality of the operation of individuation requires understanding transduction.

TRANSDUCTION

It is on the metastable, pre-individual foundation that the concept of transduction is developed as the axiomatic and ontogenetic account of how form arises. As such, transduction can be understood ontologically as an operation of ground creation that does not stipulate a prior substantive ground and asserts a priority of the transductive process to all grounds that emerge. As explained in Simondon's account of wave-particle duality, he is not concerned with any reductive move to stipulate either of the poles of this binary as primary but focuses instead on how the process of the individuation of relation establishes new dimensions and structures that are themselves fresh grounds for further operations of individuation.

Structuration is therefore understood as an ontogenetic, transductive process and being is the totality of transductive processes ever enacted. It is worth quoting Simondon's definition of transduction at length in order to enable further development of this notion:

> By transduction we mean an operation—physical, biological, mental, social—by which an activity propagates step-by-step within a given domain, and founds this propagation on a structuration of the domain that is realized from place to place: each area of the constituted structure serves as the principle and the model for the next area, as a primer for its constitution, to the extent that the modification expands progressively at the same time as the structuring operation. A crystal that, from a very small seed, grows and expands in all directions in its supersaturated mother liquid provides the most simple image of the transductive operation: each already constituted molecular layer serves as an organizing basis for the layer currently being formed. The result is an amplifying reticular structure. (Simondon, 1964: 18)

First, it should be noted that the transductive operation is axiomatic for Simondon in that it is the universal method of ontogenetic structuration; that is, it is the ontogenetic process by which all structure emerges. As such it is a positive response to the hylemorphic method. Instead of the idea that form is imposed on matter from without, the notion of transduction describes the process by which form arises in, as well as the manner by which it is amplified throughout, a domain.

Crystallization is one of Simondon's favourite examples of transduction because it clearly illustrates the restructuring of a metastable domain, the precrystalline solution that is rich in potentials, by the crossing of a threshold due to the introduction of an 'informational' seed. In this case the seed might be a piece of dust that is enough to disrupt the saturated and tense solution such that it begins to restructure (crystallize). The amplification of crystallization throughout the liquid, step-by-step, around the expanding crystal, is the transductive process.

> Becoming is no longer continuity of alteration, but concatenation of metastable states through the liberation of potential energy whose playing and existence form part of the causal regime of these states. (Simondon, 2013: 317)

In his text 'Amplification in the Process of Information' (first published in 1962 and found in the collection *Communication et information: Cours et conferences* [2010c: 159]), Simondon gives several other examples describing different modes of amplification as well as the iterative nature of transduction. The spread of an explosion along a Bickford detonating cord (such as used to trigger explosives in mining) is understood as an *additive* amplifying process as the explosion travels step-by-step along the cord at a specific rate, whereas the spread of a chemical reaction though a volume of explosive is understood as a case of *multiplicative* amplifying transfer. The spread of a forest fire is also understood as a *multiplicative* amplification in which each new tree set alight occurs by the application of an amplifying seed (a spark) causing a transition in the metastable state (dry and suitably ventilated) of the receiving tree. Given that the transductive process won't cease until the domain it is restructuring is exhausted of suitable metastable potential, it must be understood as an 'essentially positive' process whose limits are not found within itself.

However, the amplificatory nature of transduction can lead to the development of structuration through modulation. The modulatory amplification is one that occurs by 'domesticating the transductive propagation' (Simondon, 2010c: 165). Such domestication occurs by regulating the energy input into the receiving domain, thus leading to a controlling of its rate of change.

Such structuration can itself lead to a more complex level of organization when two or more modulating processes (structures) and transductive processes come into relation, leading to their becoming incorporated in an organizing systematic that resolves incompatibilities at a higher level. An example Simondon gives of this kind of organized amplification is the manner by which binocular vision results from the resolution of two disparate images.

Another essential component of Simondon's theory is that what is individuated is not just an individual but always an individual in relation to a milieu. For Simondon potential resides both in the pre-individual singularities that enable fresh transductive process as well as in the energetic continuums he calls milieus.

An individual is never fully individuated. All being has the capability to individuate further as all individuations carry with them potential energy that can be the ground for further individuations. The reason for this is that the operation of individuation does not exhaust all the potentials of the pre-individual reality from which it develops. There is a remainder of non-actualized potentiality that also emerges alongside the individual in the individuation process and that Simondon calls a milieu. Simondon describes the milieu as a 'system, a synthetic grouping of two or more levels of reality' (Simondon, 2013: 30n8) but the resolution of the disparity between these levels is never a complete resolution without remainder. If that were the case then all the potentials of the system would be made actual.

In the example of the crystal, we begin with a supersaturated solution, which is rich in potential and whose metastability is disturbed by the seed (perhaps a piece of dust) around which the transductive crystallization occurs. As a result of this process we are left not just with a crystal but also the transformed mother-water that is its milieu. There has been a restructuring of the domain. The resultant individual-milieu dyad is thus the resolution of potentials belonging to the two orders of magnitude of the seed and the solution. Given that this is a very straightforward example of amplificatory transfer, the transduction process will cease when the crystal exhausts the relevant potentials of the supersaturated mother-water.

The idea of the milieu is extremely important as it means that the problem of substantial identity is kept at bay, in that any individual is always stipulated as being incomplete (a seemingly complete crystal would continue growing if placed in another supersaturated solution), in that it always remains open to transformation from the pre-individuality in which it is situated.

Simondon partially develops the concept of the milieu from that of the field. The example of an electromagnetic field provides a good example of how the individual-milieu relationship is reciprocal and how it can lead to transformative modes of interaction productive of modulated relations. He considers the introduction of a non-magnetized piece of soft iron into a magnetic field that has been constituted by the placement of three magnets in a room. Prior to its placement in the field the iron 'does not possess this selective mode of existence, which is characterised by the existence of poles' (Simondon, 2013: 545). However, the field magnetizes the iron and in doing so the iron's magnetism has a reciprocal effect on the structure of the whole field itself—it 'becomes a citizen of the republic of the whole'. This is what Simondon describes as 'the reciprocity between the function of totality and function of element in the interior of the field'.

The individual, then, is always in relation to its milieu, which co-individuates along with it. As such the individual can never be considered as complete but always partial and in the process of individuation, the milieu always acting as a mediation between individual and world. As we will see, as organizational complexity increases, individuals hold double relationship both internal to themselves as well as to an external milieu.

Although Simondon was writing prior to the development of the sciences of complexity, chaos and self-organization, his ideas prefigure much of what these disciplines describe and also provides an ontology that enables a way of developing a metaphysical account aligned to their findings. Specifically, we can look to some of this work to help us understand the importance of the notion of the phase-shift and how Simondon complements it with his own reformulation of the notion of information.

In his trilogy of books on pattern and self-organization Philip Ball (2009a; 2009b; 2009c) studies a wide range of phenomena (physical, biological and social) in order to understand the processes that lead to the creation of recurring pattern. Ball's work is both a comprehensive cataloguing of these processes and their scientific histories as well as an attempt to investigate if the phenomena of patterning seen in one instance are transposable to phenomena in vastly different situations, in a manner similar to cybernetics' analogical method regarding organisms and machines. In that instance the drawing of such an analogy was problematic for Simondon as it was a structural analogy (resemblance) rather than an operational one.

Ball's main concern is not in developing an overarching law of pattern that can account for the broad range of spontaneous pattern-forming processes that he describes; nevertheless he does write, 'What Nature uses is not a law of Pattern but a palette of principles' (Ball, 2009c: 180). At one point in his work he does come close to developing an overarching axiomatic principle:

> Competition lies at the heart of the beauty and complexity of natural pattern formation. If the competition is too one-sided, all form disappears, and one gets either unstructured, shifting randomness, or featureless homogeneity—bland in either event. Patterns live on the edge, in a fertile borderland between these extremes where small changes can have large effects. This is, I suppose, what we are to infer from the clichéd phrase 'the edge of chaos'. Pattern appears when competing forces banish uniformity but cannot quite induce chaos. It sounds like a dangerous place to be, but it is where we have always lived. (Ball, 2009c: 183)

The similarity here to Simondon is tantalizing and significant. The key point of difference is with Ball's claim that it is competition between forces that is responsible for the emergence of pattern. In the paragraph cited it is not clear whether Ball is arguing for an ontology based on pure force or if he should be interpreted as focusing on entities that relate to one another *via* forces. While this lack of clarity is perhaps not surprising, given that Ball is not aiming to outline a philosophical ontology but just present varied examples of self-organization, it is vague enough to introduce a significant gap that could be inhabited by both substantialism or dispositional theories due to a lack of stipulation of ontological ground.

For the present, my claim is that Simondon's notion of transduction is a more consistent concept than competition to capture the impersonal, immanent processes described by Ball. By stipulating that transduction arises from pre-individual tension Simondon captures this sense of competition immanent to the process without the danger of reducing it to being between already individuated entities. That is, Simondon remains ontologically consistent by insisting that this tension is a condition *of* being, rather than a condition *existing within* it.

Another important observation made by Ball in this concluding passage from his book series is that pattern/form occurs in a situation between 'featureless homogeneity' and 'unstructured, shifting randomness'. Although this statement will resonate with anyone familiar with chaos theory it also chimes with Simondon's transduction when he asserts: 'Information is thus midway between pure randomness and absolute regularity. We can say that form, conceived as absolute regularity, as spatial and temporal, is not information but a condition of information' (Simondon, 2001: 137).

We will find time and again that Simondon is interested in undermining terms that sit at the extremes of relations (for example, form/matter, determinism/indeterminism) by uncovering what happens in the 'obscure zone' between them, which is actually where relation gives rise to those terms. The role of relations in Simondon's ontology is crucial. For Simondon a relation is not something formed between entities but emerges through the act of the resolution of a disparity of orders of magnitude in an individuation itself. That is, relation emerges from the individuation of being and is not something that is constructed between two already existing entities. As such, being can be understood as relational in the sense that its fundamental condition of falling out of phase with itself is, at the same time, productive of relations as well as relata.

Thus not only does form not arise via hylemorphism but also, unlike in Gestalt theory, form is not something that is in a state of stable equilibrium against a passive background, but is the result of the modulation and organization of processes of amplification, which are resultant of metastability, the openness of a milieu and generative of relations.

From the position of theories of self-organization the key to understanding the production of form is via the notion of phase transitions. Ball describes these as the 'largely unseen bedrock of all physics today' (Ball, 2009c: 182) and when considered in terms of form they enable the development of the idea of symmetry breaking. As Ball explains:

> The problem of creating patterns and forms that we tend to recognize as such is therefore not one of how to generate the symmetry that they often possess but of how to reduce the perfect symmetry that total randomness engenders (when considered on average), to give rise to the lower symmetry of the pattern. (Ball, 2009a: 23)

The problem then is how does form develop from the total randomness (perfect symmetry) of true thermodynamic equilibrium; in what does this symmetry breaking process consist? Ball goes to great lengths to demonstrate how symmetry can be 'spontaneously broken' and additionally that when it is, effects often differ from their causes. For example, when a liquid is heated uniformly from below it will spontaneously develop hexagonal circulating convection currents. This is the emergence of form in the liquid that doesn't reflect the nature of the force that causes it (the heating). This is what Ball calls getting 'order for free', which means 'getting order out without putting order in . . . although it is more correct to say that symmetry is being lost rather than order gained' (Ball, 2009a: 25).[3] Such symmetry-breaking phenomena are evidence for a non-hylemorphic account of individuation given that form is not imposed on matter mould-like but spontaneously emerges.

A phase transition occurs when a system shifts from one equilibrium state to another, such as when water turns to ice or, as we saw in the example above, when a liquid is heated in such a way as to produce convection rolls. These transitions occur when certain systemic thresholds are crossed that leave the system in a metastable or critical state. With water this state is reached around 0°C although, as Ball notes, 'Water can be supercooled below freezing point without turning to ice, if it is free from small particles on which ice crystals might nucleate' (Ball, 2009c: 193). This aspect of freezing water marks it out as a first-order phase transition because the transition requires a seed to initiate structural change and then propagate it through the system from that point.[4]

A second-order phase transition (critical phase transition) differs from the first-order variety in that there is no requirement for a seed to initiate the transition as it will always spontaneously occur when the system passes a certain threshold. Unlike a first-order transition there can be no delay in the transition occurring (hysterisis) due to the lack of a seed and when the transition occurs it does so simultaneously throughout the whole system. For example, the convection rolls discussed above spontaneously occur throughout the whole body of the liquid when the threshold conditions are reached without the need for a seed or some form of propagation.

Simondon places the notions of phase and phase transitions at the heart of his theory of ontogenesis. As we will see shortly he also adds a reformulated notion of information that enables the instantiation of higher levels of organization. It is clear from his use of crystallization as the paradigmatic example of transduction that this notion is directly related to that of phase transitions. Simondon's unique move comes with utilizing this concept as axiomatic for his whole ontology. What this enables him to achieve is an ontological method that underpins a coherent realism.

As Ball makes abundantly clear, phase transitions are the cornerstone of many different physical, vital and social processes. What this work on self-organization doesn't do, however, is situate these explanations of diverse phenomena ontologically. One of the features of Simondon's project is that it does just this. Although the example of crystallization is clearly a first-order transition in that it requires a seed to instigate individuation, Simondon clearly also takes into account second-order transitions as well in his descriptions of different types of amplification. The emphasis he gives to a first-order mode of amplification is due to the manner of its progression, which enables the development of a simultaneous topology, involving the development of levels, as well as a chronology of a systemic development that isn't given all at once:

> An essential characteristic of transductive propagation is the existence of a trigger threshold and a quantum character of operation, proceeding all or nothing, and implying, after each change of state, a refractory period (or recovery time) during which no informational impact is effective. (Simondon, 2010c: 164)

Another reason Simondon focuses particularly on the first-order phase-shift is that the requirement for a seed to generate a propagation across a domain is important for him so he can link this thermodynamic theory with the cybernetic notion of information. The importance of this can only be ascertained if we fully explicate Simondon's theory of levels or orders of reality and how information enables communication between them.

INFORMATION

As we have seen Simondon was critical of the uses to which cyberneticists put Shannon's information theory. In response he proposed a reformulation of the concept of information such that its primary expression is no longer as a mathematical measure of the uncertainty surrounding the communication of a message between two entities. Instead, information is descriptive of the process by which individuation occurs and as such is often used by Simondon as a description of different aspects of that process. Thus the term is sometimes used to describe the difference between levels of reality that come into contact and are productive of a state of disparity, which Simondon calls a problem for being, and that require resolution into a higher level. It's also used by Simondon to name the demand for the resolution of that problem as well as 'the arrival of a singularity establishing a communication between levels of reality' (Simondon, 1964: 130) resulting in a new dimension. Additionally, in some places information is used as the term that describes the signification that this process produces. As such, for Simondon, *information* is the term used to describe the individuation process from a number of different perspectives.

The example of crystallization is instructive. The seed on which the crystallization process depends to get started is the singularity that enables the communication between the microphysical and macrophysical levels, which are the molecular and structural crystalline levels. In this example the supersaturated solution is one level of reality, that is, in a state of metastability that with the arrival of the singularity begins a process of dephasing to another physical level, that of the crystal.[5] It is this process that Simondon names information and the mode of individuation is transductive. The process of information initiated by the mediation of a singularity and resulting in a 'communication between orders of magnitude' can be found in many other examples in nature; Simondon gives the examples of 'the stone that initiates the dune, the gravel that is the seed of an island in a river carrying alluvium' (1964: 36n1), which Ball also describes along with other examples such as the growth of cities (2009c: 65), avalanches (2009b: 104) and snowflakes (2009c: 20).

Each level is a phase of being and 'becoming is a dimension of being corresponding to a capacity of being to fall out of phase with itself, that is, to resolve itself by dephasing itself' (Simondon, 2010b: 6), a capacity that is the consequence of the metastable condition of being.

This structuration is simultaneously the creation of a corresponding chronology and topology that pertains to that system as a dimension of its individuation. For any system, spatiality and temporality are emergent from the process of the individuation of pre-individual being and subsequent individuations thereafter. The non-deterministic and discontinuous nature of transduction produces different temporalities associated with different transductive operations. Modulation also entails a further level of the control of the temporality and spatiality of individuation that becomes more complex with the production of an organization, producing what Simondon calls a thickening of time.

It is therefore a consequence of this theory of becoming that it requires a retheorization of what an individual is. It is no longer possible to understand the individual as a self-contained entity existing within an absolute Newtonian temporality. Instead,

> The physical individual must be thought of as a chrono-topological ensemble whose complex becoming is made from successive crises of individuation; the becoming of being consists in this non-coincidence of the chronology and the topology. The individuation of a physical ensemble would then be constituted by the concatenation of successive regimes of this ensemble. (Simondon, 2013: 149)

The individual, then, is never one; it is instead a series of resonating phases bringing forth a topological-chronological structure and as such is always susceptible to further informational occurrences.

It is important that we do not understand Simondon as equating being with a receptive matter or a substantial continuum on which forces/powers act. To think in terms of being in this way is to fall into the dualistic thinking of the opposition between matter and energy (such as Newton's dichotomy of bodies/forces or particle/energy, individual/field) but it is precisely these oppositions Simondon wishes to go beyond. Instead these oppositions themselves should be understood as dimensions 'which surge forth in the real when it individuates itself' (Simondon, 1964: 127). In Simondon's work we frequently find the dissolution of dualisms through a demonstration of their actually being the extremes of a specific relation and thus not being substantial.

This theory of ontogenesis is also consistent with the conceptualization of time given in Machian dynamics, that is, that time arises from the dynamic change of the universe. That is to say that in Simondon there is no account of absolute time but indeed different relations are constitutive of their own temporalities.

Another important concept regarding how we comprehend that nature of an individual, and how its development relates to groups of other individuals, while always remembering that any individual's coherence as such is always itself metastable, is that of resonance. A group of individuals can only itself be described as an individual if the milieus of the individuals are situated so that they can affect one another. If this is not possible there cannot be said to be a true grouping and the situation offers no possibility for further individuation.

For example, a number of fissionable nuclei cannot really be said to be a group, or form a level, if the fissioning of one of the nuclei cannot result in the fissioning of another. In that case each nucleus should be regarded as individual as 'each has its own chronology and fission occurs for each nucleus as if it is alone' (Simondon, 1964: 129). However, if the fissioning nuclei are situated in a resonant state such that the fissioning of one triggers the fissioning of the others, then the entire collection of nuclei can be regarded as one individual with its own chronology and topology. This not only helps us understand what an individual means for Simondon (that is, any grouping that has metastable limits that can receive information) but also how different levels can communicate. Therefore any individual is always, for Simondon, constructed from nested or imbricated levels that ultimately have arisen via individuation from the pre-individual. In this example we also see the importance of the field or milieu for delimiting the limits of an individual.

By now it should be clear that the thermodynamic notion of the phase shift is axiomatic for Simondon's informational theory of becoming, and what this entails, due to the emergence of levels this process involves, is the possibility for a range of chronologies inhering within a single individual:

> In fact, the general case [of becoming] is that of quantum thresholds of resonance: for a change occurring at one level to attain the other levels, it must be above a certain value; internal resonance is accomplished only in a discontinuous manner and with a certain delay from one level to another; the physically

individuated being is not entirely simultaneous in relation to itself. (Simondon, 1964: 126)

The specification that a threshold must be crossed before one level of reality affects another is recognizable from phase-shifts in thermodynamics. Simondon's reworking of the notion of information enables him, however, to move away from the kind of cybernetic systems thinking that was developed from an engineering perspective.[6]

Although Simondon still utilizes the vocabulary of cybernetics, with references to systems, his notion of information shifts his theorization away from the sender-receiver model, which cyberneticians such as Wiener found so useful for developing analogies between animals and machines. Instead his notion of information as ontogenetic still enables systems to condition themselves but via processes of structuration and resonance. That is, (metastable) equilibrium structures individuate but are still able to fall out of phase with themselves, either from the prompting of an internal resonance (via a singularity) or by influence coming from outside (milieu), thus leading to a further need for resolution via individuation.[7] This reformulated notion of information also changes the understanding of noise from being that which interferes with the transmission of a message to the possible indeterministic introduction of a singularity into a system. As such Simondon's theorization of what a system is means it is far more open to an outside than any other found in cybernetics.

In the early cybernetic model the notion of information corresponded to the transfer of a message from a sender to a receiver within a system that presumed the pre-establishment of entities between which messages could flow. It is important to understand that in Simondon's conceptualization of information terms do not exist prior to the operation of individuation, therefore this cybernetic model of information exchange is not yet apposite. This is not to say that Simondon holds the quantitative theory of information to be erroneous, which is certainly not the case; only that whenever we see a context in which such information is being described it necessarily depends on the priority of the operation of 'primary information' having already occurred. That is, any individuals that can send or receive information in the classical cybernetic sense must already have first been individuated from the pre-individual by a process of information in Simondon's sense. This is crucial, for to emphasize the primacy of information understood as message, as some cyberneticians arguably did,[8] is to make a mistake regarding ontological priority. It is just such a mistake that enabled some cyberneticians to equate mechanisms with organisms.

This isn't to say that embracing a Simondonian prioritization of primary information entails ruling out the possibility of highly determined mechanisms or the mathematical interpretation of some systems. However, what we tend to find is that there are a broad range of systems with varying levels of determinacy and that indeterminacy does not have to be located at the level of the already individuated information-as-message but can be found at a more fundamental ontological level.

Both total determinacy and total indeterminacy are rare in actuality. To understand why we need to understand what it would entail for a system to be either of these. For a system to be entirely deterministic would require that its becoming be entirely predictable because it could not be the site of further individuation—that is, the structure of the system would not be changed by further exchanges of energy with itself, or between different imbricated levels of that system, thus 'leaving it topologically identical to what it was in its previous states' (Simondon, 1964: 124).

A determined system, so theorized, that contained no metastability would then have the qualities of the philosophical conception of a substance, which is identical only with itself. Such a substantial kind of being would be unable to individuate as it 'would be a physical individual totally resonant in respect to itself, and consequently completely identical to itself, perfectly coherent with itself and one' (Simondon, 1964: 126). A determined system necessarily lacks internal resonance and 'we could know the becoming of this system by continuum theory, or according to the laws of large numbers, as thermodynamics does' (Simondon, 1964: 126).[9]

At the other extreme, for a system to be wholly indeterminate would require it to have 'an internal resonance so elevated that any change occurring at any particular level would immediately reverberate at all levels in the form of a change of structure' (Simondon, 1964: 126). That is to say that an indeterminate system would lack any form or structure that could be maintained over time. The becoming of such a system would result in an absence of a 'correlation between the topology and the chronology' of the system.

Complete determinism and indeterminism describe the theoretical limit cases regarding the development of individuals and it is at either end of this polarity that the conflicting traditional notions used for describing the real have been forged—'the continuous and the discontinuous, of matter and energy, of structure and operation' (Simondon, 1964: 129). What we find, however, is that systems are situated, and individuation occurs, between these polarities and involve both the continuous and discontinuous (for example, energy and particle) in such a way that an ongoing 'correlation between the chronology and topology of a system' results. By focusing on the terms that are at the poles of relation what is missed is the 'intermediary being' (Simondon, 1964: 130), which connects and supports them:

> but the operation of individuation is the active center of this relation; it is this [operation] which splits that which is unitary into aspects that are complementary for us whereas in reality they are coupled by the continuous transductive

> unity of intermediary being, which we call here internal resonance; the complementary aspects of the real are extreme aspects defining the dimensionality of the real. (Simondon, 1964: 130)

As Simondon writes, this correlation between chronology and topology operates at a level that is

> variable in function of the vicissitudes of its own becoming; a system which reacts on itself not only under the principle of entropy, by the general law of its internal energetic transformations, but also by modifying its own structure through time. (Simondon, 1964: 125)

Individuation must operate in accordance with the second law of thermodynamics, which insists that any change in a system always results in an increase of entropy. A good example of this is the formation of soap bubbles, which take the form they do because it minimizes their total energy most efficiently, which is why what is formed is known as an equilibrium structure. However, many form-giving systems are non-equilibrium structures (due to their openness to the world) and thus have a different relation to entropy:

> The thermodynamics of non-equilibrium systems is concerned not with some end point in which entropy has increased in relation to the initial state; rather, it considers the process of becoming, of how change occurs. (Ball, 2009c, 187)

It's precisely regarding how it is that change occurs, particularly that which is negentropic, that concerns both Simondon and those investigating self-organization. They are particularly concerned with how systems persist without contradicting the second law of thermodynamics, which maintains they should always be moving, due to probability, towards entropy. It must be remembered that even so-called equilibrium structures will eventually dissipate (for example, a bubble won't last forever, iron turns to rust, etc.).

This is a problem that has concerned those studying thermodynamics for some time. Ilya Prigogine's notion of dissipative structures is one important account of how negentropic structures are thermodynamically possible and also supports Simondon's overall thesis. A dissipative structure is one that is prevented from reaching thermodynamic equilibrium due to boundary conditions and therefore 'the system settles down to the state of "least dissipation"' (Prigogine, 1978: 779), this state is one in which entropy is still produced but at the most minimal level. One important conclusion Prigogine drew from this was that these processes are irreversible and therefore describe an arrow of time, which is echoed in Simondon's claims regarding chronology and topology.

Another important discovery was of systems adopting structuration occurring far from equilibrium, a discovery that itself influenced the development of the scientific study of chaos. The example of convection rolls is a good example of this in which, without means of prediction, the application of energy into a system takes it far from its equilibrium state and results in the formation of a new macroscopic order.

A further development has been Edwin Thompson Jaynes's principle that, contrary to Prigogine's theory of 'least dissipation', in non-equilibrium systems entropy tends to get maximized, that is, 'a system tends to adopt the state in which entropy is produced in the greatest rate' (Ball, 2009c: 207). If true, then because it develops via non-equilibrium systems that maximize entropy production, life in the universe can be argued to be inevitable. As Roderick Dewar describes it, 'far from equilibrium, the coexistence of ordered and dissipative regions produces and exports more entropy to the environment than a purely dissipate soup' (quoted in Ball, 2009c: 207).

Returning to Simondon we can now see how his reference to entropy in relation to an individual's becoming makes sense in thermodynamic terms, as the development of structure does not contravene the second law. Simondon's account of individuation concurs with that of dissipative structures in that they require a constant input of energy (hence his persistent references to metastability, modulation, milieu and potential energy) and the mode of their becoming is via phase-shifts (transduction) usually occurring between disparate levels. That a system (or quasi-system, to use Simondon's preferred term) is dissipative requires it to be not just open to an environment but also intimately connected to it—aligning with Simondon's assertion that it is never just an individual that comes into being alone but an individual-milieu dyad.

Furthermore this makes clear why Simondon asserts that a 'physical being . . . must be considered as more than a unity and more than an identity, rich in potentials; the individual is in the process of individuation starting from a pre-individual reality that sub-tends it; the perfect individual, fully individuated, substantial, impoverished and empty of its potential, is an abstraction' (Simondon, 1964: 126).

The nature of individuation disqualifies the thinking of an individual apart from that which subtends its development (thus why it is more than a unity or identity) and this is why, for Simondon, a relation between two individuals is not something established after the individuation process but as part of this process. It is also why the individual is always considered as having only 'a relative coherence in relation to itself, a relative unity and a relative identity' (Simondon, 1964: 126).

CONCLUSION

Having covered the main concepts of Simondon's ontology, it is worth summarizing it in order to draw out some key aspects as well as highlight some of the implications that will require development in the remaining chapters.

As discussed above, Simondon develops an informational account of ontogenetic individuation that he applies to all levels of being. The progression of individuation consists of the shifting of phases. Simondon interprets this process of phase-shifting into a universal informational axiomatic for the creation of new levels, which are considered both as new grounds on which further becoming may occur and as new dimensions of being. The claim then is that at any level of being there is some level of metastable equilibrium that can be broken and lead to further individuation, and that the mode by which this occurs is transductive. Therefore transduction is the barest description of the immanent operation of individuation that is being.

To outline Simondon's core argumentation for the priority of individuation: for any individual an account is required for how that individual came to be individuated. Such a demand is also required for any substantial account of what constitutes being and, what's more, individuation is not possible from any substance that is identical with itself (that is, that contains no metastability); therefore if there is individuation this must be because being itself is nothing other than individuation from which further individuation proceeds.

What this provides us with is a new understanding of nature. Crucially, this ontology avoids the pincers of both mechanism and vitalism, as well as doing away with dualism. This is a philosophy of nature that is founded on the operation of a pre-individual monism:

> Nature is not the opposite of Man, but the first phase of being, the second being the opposition of the individual and the milieu, complement to the individual in relation to the whole. (Simondon, 1989: 196)

Although Simondon's is a monist ontology, we have seen that variation and difference emerge from the creation and resolutions of disparity as well as modulation and organization. Going further, he actually distinguishes three regimes of individuation that mark out qualitatively different modes; these are the physical, the vital and the psychic and collective.

We can now more clearly understand why there is a requirement for an allagmatic epistemology, given that individuation operates across all domains. Specifically, this 'new method' offers an alternative to philosophical thinking that relies on identity, that is, that considers already individuated terms and the relations that are thought to hold between them. Simondon maintains that thought that relies for its method on the law of identity and the law of the excluded middle 'is too restrictive' if being is fundamentally pre-individual in nature.

The transductive method requires us to understand individuals as neither fully determined nor undetermined. An individual, therefore, always occurs in a state that is never wholly stable or unstable. This ontology requires the development of a new method that utilizes this transductive axiomatic. That is, that does not rely on constituted terms (such as we find in the syllogism, for example) or negation and synthesis (such as we find in dialectics), as for Simondon all individuals, such as they are, are positive.

> We could give the name of allagmatic to such a genetic method that aims to seize individuated beings as the development of a singularity which unites to a middle order of magnitude the global energy and material conditions. (Simondon, 2013: 82)

Transduction and the production of new levels also provoke questions regarding the nature of causality, especially in far-from-equilibrium situations. In particular it demands an account regarding both the role of indeterminism in such phenomenon as well as the fact that effects are different from causes; that is, there is often nothing in the cause that indicates what its effect might be. This is, of course, one of the founding tenets of chaos theory when describing open systems.

It is perhaps because of this that Simondon places more emphasis on first-order phase transitions in order to illustrate his transductive theory rather than the second-order variety. With the first-order variety of transition the jump between levels is more gradual and requires a singularity (an informational seed) to cause the initial destabilization of the level. This processual aspect allows Simondon to develop a relation between levels via the notions of resonance, singularity, information and the individual-milieu environment.

One of the roles of the singularity in this process is to enable recursive causality in systems as the singularity can be a point at which a system can modulate itself in a way similar to that found in cybernetic accounts of self-regulation. As we will see this underpins the key concept of signification for Simondon. There are different mechanisms at work here that need to be distinguished: the resonance that adheres between individuals that enables the attainment of individuality as well as the mechanism that enables the breaking of the equilibrium that inheres between and within such individuals. This identification will help us elaborate the tension at the heart of Simondon's ontology between regulation and invention.

NOTES

1. Although in his doctoral thesis and in the corresponding paragraph in *L'individu et sa genese physico-biologique* (1964: 286) Simondon claims that the hypothesis of the pre-individual 'is derived from a certain number of schemas of thought borrowed from domains of physics, biology, technology' (2013: 317), it is only science that is mentioned in the same passage in the later published *L'individuation psychique et collective*: 'it [pre-individual] is derived from a certain number of schemas of scientific thought, particularly the thought of physics' (1989: 232).

2. Simondon makes several references to the apeiron as described by Anaximander throughout his writing, which he relates to his notions of both the transindividual and the pre-individual. In his text 'Histoire de la notion d'individu' he describes the apeiron as 'the infinite and indefinite' and links it to the notion of homogeneity, especially regarding some natural patterns. For example, 'Natural patterns like cloud, air, water make tangible this link of unity and homogeneity' (Simondon, 2013: 358). In *L'individuation psychique et collective* Simondon connects the notion more definitively to that of the pre-individual: 'One could name nature that pre-individual reality that the individual carries with it, by seeking to find in the word nature the significance that the pre-socratic philosophers gave it: the Ionian physiologists found the origin of all species of being, anterior to individuation: nature is the reality of the possible, under this kind of apeiron Anaximander created all forms of individuation: Nature is not the opposite of Man, but the first phase of being, the second being the opposition of the individual and the milieu, complement to the individual in relation to the whole. According to the hypothesis presented here the apeiron would still remain in the individual, like a crystal which retains its water-mother, and that charge of apeiron would enable a going towards a second individuation' (Simondon, 1989: 196).

3. A connection can be made here to von Foerster's important essay for neo-cybernetics, 'On Self-Organising Systems and Their Environments' (von Foerster, 2003: 1–19), in which he creates the thought experiment of a bag filled with magnetic blocks that upon shaking become more ordered. This is a cybernetic twist on the classic thermodynamic thought experiment of Maxwell's daemon but in this case increased order in a system is created without order (only energy) being put into the system. Von Foerster calls this his 'order from noise' principle. As with Simondon, von Foerster's example also requires that the system is engaged with an environment from which it receives energy.

4. Ball describes many other occurrences of this kind of structural propagation in naturally occurring phenomena that can be understood via transductive amplification, modulation and organization. For example, the way that butterfly wings (Ball, 2009a: 193) are patterned by a chemical reaction-diffusion system that consists of *determination waves* emanating from morphogen generating points that are channelled and modulated by the veins in the butterfly wings. Transduction could also be likened to reaction-diffusion phenomena that have been analysed at various levels from the chemical, from Belousov's 'oscillating reaction' in glycolysis (Ball, 2009a: 111) to the ecological in Lotka's study of predator-prey relations, which oscillate over time (Ball, 2009a: 117).

5. It is in precisely this way that Simondon's notion of information corresponds to Bateson's formula that information is 'the difference that makes a difference' (Bateson, 2000: 459).

6. As discussed in the previous chapter, both Wiener and Ashby developed a notion of information that enabled animals and machines to be understood as systemically analogous. As we saw, Simondon's account of primary information as individuation counters this perspective.

7. Another illustrative example is the creation of snowflakes (Ball, 2009c: 14–26), which involve both the equilibrium breaking operation of a singularity as well as progressive dephasing caused by environmental factors. Thus the singularity of the water molecule breaks the equilibrium of the surrounding atmosphere, the molecular shape of the molecule determining the six-branched structure of the snowflake via *anisotropy*, although the exact final shapes the branches will take is indeterminate because it depends on the variations in the conditions that the snowflake encounters as it falls. There is a constant feedback between structure and environment that informs the individuation process that occurs after the initial symmetry breaking.

8. As discussed in the previous chapter, there are times when Wiener and Ashby do this. Additionally some microbiology focusing on the importance of genes could also be understood to claim this, such as Dawkins's *The Selfish Gene.*

9. The law of large numbers is a thermodynamical law that utilizes probability to make predictions regarding the likely states of systems. By saying that we could predict the becoming of a system in such a way Simondon is indicating that the system is effectively closed.

Chapter Three

Vital Individuation

In chapter 2 I gave an exposition of Simondon's theory of physical individuation by way of a comparison with some recent work in complexity. My aim in doing so was to not only illustrate Simondon's theory through some contemporary work but to demonstrate how it could help bring some ontological coherence to that work.

In this chapter I will undertake a similar exposition of Simondon's description of vital individuation, this time in relation to Stuart Kauffman's work in biology and complexity. I hope to demonstrate that although Kauffman's biological account of the emergence of life is consistent with Simondon's, and will therefore help us explain it, that some of the philosophical implications he draws are problematic and that Simondon is useful in identifying a residual Aristotelianism.

What Simondon proposes is a thoroughgoing transformation of the understanding of individuals not as constituted by different types of substance but as the recursive interweaving of three different regimes of individuation. Although the informational operation of individuation is primary for Simondon, he is clear that there are different 'forms, modes and degrees of individuation' (Simondon, 1964: 16). The three regimes of individuation that Simondon describes are those of the physical, vital and psychosocial. As we have already seen, the prioritization of the transductive operation of individuation enables Simondon to move beyond hylemorphism and establish an ontogenetic approach replete with fresh concepts for description:

> The separation, the layering, the relationship of these domains are seen as aspects of individuation according to its different modalities; the notions of substance, form, matter, are substituted with more fundamental notions of primary information, internal resonance, potential energy, orders of magnitude. (Simondon, 1964: 17)

An initial task for a chapter focusing on the regime of vital individuation is to ask how this regime is distinguished from that of physical individuation. It must be made clear that the difference between these two regimes is not a case of their being constituted by a different substance. All individuation for Simondon emerges from the pre-individual and as such his account of how the living is individuated must also be tracked in the same ontogenetic manner with which we described physical individuation. I will attempt to address the discernment of vital individuation from that of the physical by working through the four key interrelated concepts Simondon uses to describe this regime: level, modality, topology and chronology.

LEVEL

The notion of orders of magnitude complements Simondon's theory that individuals are constituted though imbricated levels. An individual emerges from the bringing into relation of different orders of magnitude such that the individual can be understood as a relation. An example Simondon gives is of a plant that, when described in terms of orders of magnitudes, is the bringing into relation of the cosmic order (particularly sunlight) and that of the soil. The plant individuates on a new order of magnitude as a relation between these two orders.

Simondon holds that vital individuation occurs at the level of the *macromolecular.* This level of the real is situated between the microphysical and the macrophysical. He describes the microphysical as that which is of an 'inferior order of magnitude' to the macromolecular, and as being 'in fact neither physical nor vital, but pre-physical and pre-vital' (Simondon, 1964: 131). Thus for Simondon we can only talk about physical and vital individuals as occurring above this level, on the macromolecular level where individuation can produce 'a crystal or a mass of protoplasmic matter' (Simondon, 1964: 131).

The distinction between the living and non-living is to be found at this macromolecular level of organic chemistry and neither at the pre-physical level nor at the level of more complex individuals. As such Simondon is arguing that for an entity to be understood as vital does not require it to have a sophisticated organization; even a unicellular organism is considered as living. Rather, the vital is established as a certain mode of individuation occurring at the macromolecular level of being.

MODALITY

Simondon's ontology resists any kind of substantialism. He does hold, however, that there are regimes of individuation that are described by differences in the modality of individuation. These modalities must not be mistaken for different substances, however.

In the process of crystallization, which is Simondon's paradigmatic example of physical individuation, individuation occurs in 'a manner of *instantaneous* settlement, quantum, sudden and definitive, leaving after it a duality of the milieu and the individual' (Simondon, 2013: 27). Such physical individuation does not perpetuate itself beyond the initial resolution of the original disparity, which results in the formation of a stable and unified crystal and what it leaves of the mother-water from which it transductively emerged.

Although vital individuation also develops from the pre-individual, it does so in a manner that 'keeps it in a permanent activity of individuation, it is not just the result of individuation, as crystal or molecule, but a theatre of individuation' (Simondon, 2013: 27). Unlike the individuation of a crystal, in which once individuation occurs it exhausts the pre-individuality of its milieu, vital individuation continues through the establishment of a sustained metastability, 'which is a condition of life' (Simondon, 2013: 27). The establishment of a continuous mode of individuation, which Simondon argues characterizes the vital, requires that the individuating system doesn't exhaust itself in just one moment of information but that it's able to receive and self-limit repeated informational singularities, thereby extending the activity of individuation indefinitely. As such Simondon recognizes that physical and vital individuation both emerge at the same macromolecular level and one does not build on the other:

> We would assume that vital individuation does not come after physicochemical individuation, but during this individuation, prior to its completion, by suspending it at the moment when it has not attained its stable equilibrium, and rendering it capable of extending and propagating itself prior to the iteration of a perfect structure capable only of repeating what would conserve in the living individual something of the pre-individual tension, of the active communication, in the form of internal resonance, between the extremes orders of magnitude. (Simondon, 1964: 132)

It is thus an extension and prolongation of the individuation process at the level of the pre-individual that enables the establishment of a mode of individuation that is vital. Both physical and vital individuations emerge from the macromolecular level of the physico-chemical (as such they are both physical) but the vital describes a modality of individuation that remains 'in abeyance long enough at its origin' (Simondon, 2013: 233). Simondon also describes this abeyance of individuation, which denotes the vital mode of individuation, as neotenic: by this he means that there is in vital individuation a slowing down or modulation in the speed and rhythm of the individuation process, which means it 'retains and dilates the earliest phase of physical individuation—such that the vital would be the physical in suspense, slowed in its processes and indefinitely dilated' (Simondon, 1964: 133).

This slowing of the individuation process and variation of its rhythm provides further opportunities for the reception of information to be made available, which can enable the system to both stabilize and organize.

TOPOLOGY

Having established that vital individuation involves a certain prolongation and modal variation of physico-chemical individuation at the macromolecular level, it is important to recognize that the informational opportunities that this enables are topological as well as chronological. As Simondon states,

> The bodies of organic chemistry do not provide with them a different topology to that of physical relations and habitual energy. However, the topological condition is perhaps primordial in the living qua living. (Simondon, 1964: 258)

Simondon makes a number of claims regarding the central importance of topology for the 'essence of the living being' (Simondon, 1964: 259). For instance he claims that it is the maintenance of 'certain topological conditions' rather than just purely energetic or structural ones that characterize the living. Additionally, such topological conditions are ontogenetically emergent with vital individuation and so cannot be 'known on the basis of physics and chemistry' (Simondon, 1964: 259).

The topological characteristic that Simondon views as foundational for all vital function is that of the permeable cell membrane because its establishment is simultaneously also the creation of regions of interiority and exteriority. This topological development is important, as it enables the instantiation of the 'polarized, asymmetric character of cellular permeability' (Simondon, 1964: 259). This polarity establishes the membrane as a barrier that allows certain entities (via centripetal or centrifugal motion) into the interior of the cell from the exterior while also blocking the passage between the two. It is this *selectivity* that Simondon argues 'makes the living alive at every moment' and by which it maintains a 'milieu of interiority' (Simondon, 1964: 260).

At the level of the unicellular organism such selectivity may be relatively simple but it still gives the organism 'a direction inwards and a direction outwards' (Simondon, 1964: 260), which Simondon argues provides the foundation for agency.

Returning once again to the example of crystallization we can see how the topological characteristics of individuation help establish the difference between physical and vital beings. The individuation of the crystal only occurs on the outer molecular layer of the crystalline formation; its interior remains as an inert historical record of previous individuating activity. However, in the case of even the most rudimentary cell, although the boundary plays the important role of selection and partitioning, it also enables the maintenance of an interiority that is both a 'condensation' of past activity and also the maintenance of an internal resonance throughout the topology of the whole cell. This resonance is both the source and receptor of constant communicative activity between the interior and exterior, as well as between structures and operations solely on the interior, thus enabling further organizational and informational activity.

> There exists within the being a more complete regime of internal resonance requiring permanent communication and maintaining a metastability that is the precondition of life. (Simondon, 1992: 305)

The topology of the cell enables the ongoing production of disparities (thermodynamic, informational and topological) in need of transductive resolution, which maintain the cell in a condition of metastability.[1] Thus when Simondon states that 'the living lives at the limit of itself, on its limit' (Simondon, 1964: 260), we can interpret this as meaning that the cell is living at both its topological limit but also its energetic and informational limit in the sense of at the edge of chaos (the critical limit between complete determinism and complete indeterminism). This account of the vital is therefore not describable by vitalism or mechanism, as it neither requires a vital principle to sustain it nor is it wholly deterministic.

Although crystallization may be given as the paradigmatic example of the transductive operation in that it tracks this operation from initial disparity to resolution, in the case of vital individuation the operation does not resolve, for in doing so it would result in the death of the organism. Vital individuation requires an ongoing metastable tension and the need for further problems requiring solution. As we have seen, the maintenance of such a situation also requires a topological organization that supports its operation. Necessarily this topological operation requires the establishment of interior and exterior milieus that sustain informational activity at various levels. As organisms become more topologically sophisticated, the level of complexity increases between the varied interior and exterior milieus that develop.

Given the emphasis on the primordial importance of topology for the emergence of the vital Simondon proposes that all organisms, however complex, can be classified not just by the level of differentiation they demonstrate but also by the levels of interiority and exteriority by which they are organized and how these are mediated. Such classification points to the different levels of topological complexity that exist in the structural organization of various organisms, with the more complex involving more opportunities for informational mediation between structures and hence the development of higher-order functional structures such as central nervous systems.

Thus the various differentiated systems that constitute an organism should primarily be understood topographically; that is, as an organized enfolding sustaining various interiorities and exteriorities that maintain relations and that have developed specific functions to aid the maintenance of the metastable resonance across the organism. Simondon calls those interiorities that contain intermediary interiorities (digestive, nervous, endocrine, etc.) and that are in communication medial interiorities. One example he gives is of the intestines:

> Depending on the topology of the living organism, the interior of the intestine is in fact exterior to the organism, although it accomplishes in this space a certain number of transformations conditioned and controlled by organic functions; this space is an annexed exteriority. (Simondon, 1964: 260)

The understanding of the organism as a topological organization in the service of the self-maintenance of a metastability has important philosophical implications. First, such an understanding disqualifies the explanatory power of a reductionist approach such as atomism or mechanical materialism when describing organisms. The fundamental unit here is the topological structure of the cell the operations of which are strongly emergent from the atomic level. For reasons given above recourse to explanations using physics or chemistry are also insufficient. Although the development and operation of organisms do not contravene physical or chemical laws, their topological and chronological characteristics cannot be reduced to either.

This doesn't mean that we should attempt to understand organisms holistically. By doing so we risk 'privileging the organization of integration and differentiation'. Evidently such organization is extremely important for any organism but Simondon's point is that it is dependent on the primary topological structure, which involves a metastable asymmetry between interiority and exteriority. Without this structure such integration and differentiation would be impossible. Simondon maintains that however complex and organized the organism is it is this topological structure that remains primary and that 'living individuation must be thought according to topological schemes' (Simondon, 1964: 262).

Such an understanding also rules out any equivalence of organisms with machines. Although I covered the inadequacy of such equivalence in the discussion of cybernetics, it is worth noting that Simondon's topological account of vital individuation provides another way in which organisms differ from technical objects. As we'll see in the forthcoming account of technology, Simondon will give an account of these involving the central role of the integration of differentiated structures (concretization); however, this process does not occur in the service of maintaining a topological metastability and so differentiates machines from organisms.

ENVIRONMENT

Another important aspect of Simondon's theorization of the vital is regarding the relation of the bounded topological structure of the organism to its environment. What is at stake here is the extent to which the topologically defined organism is closed to its environment and the implications this has for both its autonomy and the possibility for further differentiation.

It was these concerns that engaged a number of second-order cyberneticists in the development of systems theory and for which the work of Maturana and Varela is particularly pertinent. It is with their distinction of allopoietic systems with autopoietic systems that the poles of the problem were initially defined. An allopoietic system is an organization that produces something different from the organization itself. Contrasted with these are autopoietic systems, which are organizations whose operations are recursive and self-maintaining in that they self-reproduce the same processes that produced them. Additionally, through doing so, the system also constitutes itself spatially as a unity. Thus a cell is understood to be autopoietic in that its operation recursively reproduces the same set of processes that produced it and in doing so constitutes itself as a unity with a certain amount of autonomy. As a unity the autopoietic system is understood to be thermodynamically open to its environment but its operation is closed to informational exchange. The system must be thermodynamically open to be able to energetically sustain its operation; however, its boundary is required to insulate it from informational disturbance from the environment that would be disruptive of its operation. This condition is also known as 'operational closure'. This closure also enables the reduction of the influence of the environment on the system to one of signification such that the system itself constructs a meaningful order out of the noise of the environment. It is via this signification that a system is understood to be structurally coupled to its environment.

Although such theorization gives the system a high level of autonomy, the problem remains regarding how it comes to develop structurally without disruption to this required closure. As such this conception trivializes the operation of the environment in relation to the system. It is in this regard that Simondon's theorization of individuation offers a novel way to think about the openness of systems. Where in the case of autopoiesis the coupling of the system occurs at a single level of magnitude, for Simondon the system maintains an ongoing relation to the broader environment (pre-individual) as a whole. That is to say that the individual 'is sustained by a double relationship' (Simondon, 1980: 54), first with its milieu, which exists on the same level of magnitude as it does, and second with the wider metastability of the pre-individual. As Mark Hansen explains in his use of Simondon to develop his notion of the System-Environment Hybrid:

> For Simondon, in short, it is not simply the global perspective *of the organism*—a perspective tied to the organisms specification of a world—that informs the bootstrapping of identity from level to level; rather, the upward spiral of individuation is driven by two important conditions: the nonidentification of individuation with any form of individual (physical, biological, psychic, or collective) and the coupling of individuation with the entire environment as a source of 'pre-individual', 'metastable' potential. (Hansen, 2009: 134)

As such the pre-individual overflows and remains independent of the reduction that the system understands as its environment. Simondon shifts the focus from just being epistemological, which was the domain of interest for many of the second-order cyberneticists, to the ontological understanding of systems as including individual, milieu and pre-individual. It is this ontologizing of information as an individuating process that is what makes Simondon's approach so radical.

CHRONOLOGY

The topological character of the vital ensures that there is a resonance that holds throughout the interiority of the organism. This resonance enables the entirety of the organism's interior to be in contact with that which is exterior. As such any disturbance on the boundary of the organism can affect the interiority of the organism by disturbing this resonance. This means that the whole of the interiority of the organism is in effect in contact with the exterior 'without delay' enabling the sense that the organism exists in a temporal *present*. The mode of this chronological present is not constituted in 'the form of time in Physics', that is, as a linear series of punctual moments, but is specific to the topology of the particular organism, leading Simondon to claim that 'it would be necessary to define, in addition to a topology of the living, a chronology of the living associated with that topology' (Simondon, 1964: 264).

The chronology of an organism is dependent on its topology and in particular how this topology organizes the past of the interior milieu in opposition to the futurity it receives from its relation to the exterior milieu. As such this temporality is different from that of its individuation. There develops another chronology at a higher level to that of its vital individuation.

The interiority of the living individual is a condensation of its past in that it is also a physical memory of the selections that have taken place on the interior of the outer boundary. For Simondon the present of the living organism is the metastable transductive relation that is maintained between the past as constituted by the organism's interiority with the future, which is that which confronts the organism as its exterior milieu.

The fundamental chronology of the organism is thus conditioned by its topology and co-emergent with it. This entails that there must be as many kinds of chronology possible for the living as there are possible topological structures:

> Continuity is one of the chronological schemes possible, but it is not the only one; the schemes of discontinuity, of contiguity, of envelopment, can be defined in chronology as in topology. (Simondon, 1964: 264)

With this realization of the relation of the chronological with the topological Simondon identifies a primary dimensionality of the organism. This combination of a fundamental dimensionality from which higher order sensibility may later emerge places Simondon in the position of getting beyond subject-object dualism. Although obviously not denying that the subject-object schema may emerge, Simondon's ontogenetic approach ensures that it is not misconstrued as a starting place for thinking the organism.

STUART KAUFFMAN

In this section, and in the interest of aligning Simondon's thinking with contemporary science, I want to look briefly at the work of the biologist Stuart Kauffman. This will continue the connection I began in the last chapter between Simondon and contemporary engagements with complexity. However, another purpose is also to indicate how Simondon can help root out some residual Aristotelianism in modern scientific accounts. In doing so I hope to thereby demonstrate the significance of undertaking a fully ontogenetic account.

Like Simondon, one of Kauffman's primary concerns is to understand the emergence of the vital. Kauffman also proposes that it is at the macromolecular level that the development of life must occur. His reasoning for this is similar to Simondon's in that he also holds that we must think the organism from its initial organization; for Kauffman this occurs at the macromolecular level via the construction of autocatalytic dissipative structures.

Such structures must be bounded and thus he also stipulates closure as a necessary requirement for the kind of 'propagating organisation of process' that constitutes life. He writes: 'This closure is a new state of matter—the living state' (Kauffman, 2010: 93). As such Kauffman shares Simondon's view that the topology and chronology of the cellular structure is a requirement for life.

The evolution of such structures requires a kind of bootstrapping process whereby cell membranes act to constrain the release of energy from thermodynamic work-cycles while also being the result of just such processes. That is to say that the cell membrane develops as a result of chemical reactions and once constructed provides an environment that is both far from equilibrium (Kauffman, 2010: 47) and bounded in a manner that it constrains the work that continues to be done (Kauffman, 2010: 93).

The concept of work is central to Kauffman's account. He uses the Carnot thermodynamic heat engine to demonstrate the notion of a work-cycle. The operation of the engine combines spontaneous and non-spontaneous processes. A non-spontaneous process is one that requires an intervention from outside the system. This required mix of spontaneous and non-spontaneous processes for the completion of a cycle means that the cycle is a non-equilibrium concept, as at some point in the cycle the system must be *forced* away from equilibrium in order to begin again.

Additionally, what can be taken from the example of the Carnot engine is that work is not just 'force acting through a distance' but also 'the constrained release of energy into a few degrees of freedom' (Kauffman, 2010: 90). To create work it is thus necessary to constrain the release of energy, and the constraints required to do so (for example, piston/cylinder) require work for their production: 'No constraints, no work. No work, no constraints' (Kauffman, 2010: 90).

Even for the most basic cell one of the roles of the cell membrane is to act as a constraint in which to maintain the disequilibrium between its interiority and exteriority as well as to constrain and modulate the chemical-based work-cycles that occur within. Such constraint enables cells to organize in such a way as to produce further constraints with which to organize work further. This reciprocal activity of work and organization is an example of what Kauffman calls a 'self-propagating organization of processes' (Kauffman, 2010: 91).

Such *self-propagating organization of processes* emerges at the molecular level via a process involving what Kauffman calls collectively autocatalytic sets. This involves the spontaneous formation of sets of chemicals that mutually catalyse each other's reproduction and that can also form chemical reaction networks with the ability to self-reproduce. The chemistry involved in this process need not concern us here, except to state that this process enables a way to theorize the emergence and self-reproduction of more complex macromolecules.

Already there are a number of ways in which Kauffman's account is consistent with Simondon's theory of vital individuation. Like Simondon he recognizes as fundamental the topological closure of the cell for the organizing process of resolving problems related to structuration. Additionally, he offers by way of his theory of autocatalytic sets a chemical description of the kind of ontogenetic self-propagating process described by Simondon as the modality of 'perpetual individuation, which is life itself' (Simondon, 1989: 15). The requirement of constraints by which the system can then organize further corresponds to Simondon's description of modulation leading to organization.

Just as we saw in the example of the Carnot engine, the topology of the system acts as a constraint, which can organize the '*kinetic behaviour of the chemicals of which it is made*' (Kauffman, 2010: 51). Such topological constraints therefore play a causal role in the self-organizing process of the living system:

> Thus, the emergence in the universe of collectively autocatalytic, evolved cells and their 'topology' of organisation of kinetic-controlled process is ontologically emergent, and the same topology of kinetic control of the 'whole' is partially causal in constraining the kinetic behaviour of the parts. (Kauffman, 2010: 58)

With this Kauffman returns us to Kant's problem regarding the relationship of the parts to the whole of the organism with the claim that the whole effects a downward causation on the parts just as much as the parts are responsible for affecting the whole. Thus despite the similarities to Simondon noted above, at this point Kauffman's account is perhaps closer to that of autopoietic systems given its focus on bounded recursion.

One of Kauffman's aims in his books is to question the limitations of reductionism and establish a realism of entities that exist at levels higher than the molecular. One example he uses is that of the heart. If we were to give a reductionist account of the heart, that is, an explanation of it only using physics and chemistry, then there are some key properties of it that would be missed. That is, although we may be able to account for its molecular constitution and even its mechanics, Kauffman maintains that we wouldn't be able to describe how the heart came into existence, explain that its function is to pump blood or that it is a causal entity in its own right.

In short his argument comes down to the point that neither physicists nor chemists are able to predict the evolution of the heart because the evolution of organisms involves processes that, although they don't contravene physical or chemical laws, are not reducible to them because they involve processes of emergence that are 'partially lawless and ceaselessly creative'.

To make his argument Kaufmann employs the notion of Darwinian preadaptation. This holds that an entity such as an organ 'could have causal features that were not the *function* of the organ and had no selective significance in its normal environment' (Kauffman, 2010: 131). However, when in another environment these entities may indeed have significance and introduce a 'novel functionality' into the environment. The noise hearts make, for example, is not part of their function but it is nonetheless a real causal property.

Another illustration of the unpredictability of preadaptation is the evolution of the swim bladder from the lungs of early fish. Kauffman claims that such an evolution could not have been predicted. Given the unpredictability of such preadaptation, Kauffman goes on to make the stronger claim that the evolution of the biosphere is radically non-predictable because it cannot be possible 'ahead of time, or finitely [to] prestate, all possible Darwinian preadaptations of all species alive today' (Kauffman, 2010: 132).

This argument introduces the need to differentiate between weak (epistemological) and strong (ontological) emergence. Weak emergence describes what happens when a phenomenon emerges from a lower level in an unexpected and surprising manner. It is regarded as epistemic emergence due to the impossibility of predicting its emergence given knowledge of the lower levels from which it emerged. Strong emergence describes those phenomena that aren't merely unexpected epistemologically but that are instances of the production of ontologically novel phenomena that are 'not deducible even in principle from truths of lower level domain' (Mumford and Anjum, 2011: 92). This adds the further claim that the emergent phenomenon is real in the stronger sense that its causal powers cannot be reduced to activity at lower levels.

Kauffman's claim is that the preadaptations found in the biosphere are both epistemologically and ontologically emergent. This is not just to make the claim that one cannot deduce or infer the emergent phenomenon from that of lower levels (epistemological emergence) but the stronger claim that such higher-level phenomenon are real entities in their own right and are operations that can't be reduced to entities or activity at a lower level.

Such a claim has obviously damaging consequences for the ubiquity of reductionist science but doesn't mean that all higher-level phenomena should be understood as indeterminate. Kauffman's view is similar to that developed in the previous chapter in that causality for him occurs somewhere between determinism and indeterminism, that is, that ontological emergence in the biosphere is 'ontologically partially lawless and ceaselessly creative' (Kauffman, 2010: 36). So to return to the example of the heart, 'it is epistemologically emergent because we cannot deduce it from physics alone' (Kauffman, 2010: 86) and it is ontologically emergent because the laws by which it did come to be (heritable variation and natural selection) are not just irreducible but apply to that entity at the ontological level at which it exists. That is, just as we saw in the last chapter, there are causal powers that are owned by higher-level entities that just aren't reducible to the lower-level parts from which they are constructed.

Although in arguing for strong emergence Kauffman is close to Simondon, we must note a distinction in how such emergence is theorized by these thinkers as it highlights some insufficiencies in the philosophical entailments Kauffman claims. In stipulating preadaptation as the principle by which such emergence occurs, he is foreclosing the strength of this emergence as by doing so he is restricting it to the actualization of unrealized possibilities that are in some way *already present* in the actual. Thus in the examples of the heart and the swim bladder we can question if these aren't really cases of epistemological emergence as the developments involve the recognition of new *functionality* that these already have but aren't demonstrating due to their environmental context. The changes thus only involve operations at the same level of magnitude. This is different from the stronger claim that Simondon makes for transductive operations occurring across different levels and that can lead to truly novel structures. That is to say that Kauffman's account surely entails that as preadaptations these developments point just to functional changes due to their placement in a different environment rather than the stronger thesis that the environment is causally implicated in an operation of radical creation—that is, the invention of something completely new and not just adapted.

Kauffman argues that a key reason that the evolution of organisms and open systems is so unpredictable is that they are partially lawless in that no natural law governs their emergence. The reason for this is that we cannot know the space of possibilities (state space) for such an emergence prior to its occurrence. The range of possible causal influences in any situation is simply too massive to be able to predict an outcome. However, both of these points is questionable: by arguing for the principle of preadaptation isn't he specifying a 'natural' law for governing this emergence and in specifying that there is a space of possibilities, however vast, isn't he admitting this is an epistemological issue regarding the knowledge of that space?

These points can be demonstrated using another example of Darwinian preadaptation that Kauffman gives. It is a scientifically accepted fact that the three bones of the human middle ear evolved from the jawbones of early fish. Kauffman argues that it would be impossible to predict such an evolution because it's just not possible to prestate beforehand which of the many functional aspects of the entities involved would be relevant. The combinatorial possibilities of all possible environmental situations are impossibly vast. Because seemingly non-functional properties of entities (such as the sound of the heart, the distance apart of a fish's jawbones) all have real causal properties, this makes any situation's state of possibility too large to prestate. For Kauffman this kind of emergence leads to the claim that the universe is radically creative. However, this conclusion does not follow, as he is attempting to make the leap from the weaker epistemological emergence to a stronger ontological emergence.

These problems can also be detected in another important concept that Kauffman develops to augment his argument, which is that of the 'adjacent possible'. He initially develops this concept in relation to organic molecules. First, he asks the reader to

> consider a reaction graph with N molecular species, polymer sequences of A, and B monomers of diverse lengths. Call this initial N the actual. . . . The adjacent possible in relation to the actual so defined is 'the set of new molecular species reachable in a single-reaction step from the actual.' (Kauffman, 2010: 64)

The actual and the adjacent possibles are constantly changing. As the actual expands into its adjacent possible, then so too does the adjacent possible itself change as new chemical reactions are made possible.

Kauffman is clear that the concept of the adjacent possible does not refer just to organic chemistry but also applies to other real entities such as economics, the organic evolution of the biosphere, the evolution of autocatalytic sets, technological development and human history: 'The adjacent possible is real. We are invading it much of the time. . . . *History enters* when the space of the possible is vastly larger than the space of the actual. At these levels of complexity, the evolution of the universe is vastly nonrepeating, hence, vastly nonergodic' (Kauffman, 2010: 123).

However, this argument also suffers from the same criticism as that made to the examples above, which is that the creativity Kauffman is describing involves the actualization of adjacent possibles that in some sense already exist in the sense that they are already possible. As such all that differentiates these possibilities from actualization is their becoming actual. But surely strong emergence means more than just actualizing non-actualized possibilities?

Additionally this account of creativity entails that for any creation to occur there must already be some actual entity such that it may have adjacent possibles that can themselves become actual. Without such an actual entity there can be no possibles to actualize; however, the only way an actual entity can become actual in the first place is through first being possible. This therefore requires the contradictory need for a possibility that is not adjacent to an actuality.

Such a theory describes the kind of problems that Simondon's genetic theory of individuation sets out to overcome. That is to say that it lacks an account of the individuation of the entities under discussion in the first place but assumes them as already individuated. As such Kauffman's arguments rest on an implicit Aristotelianism, as it presupposes some kind of enduring actuality by which possibility can be adjacent.

With his theory of individuation Simondon attempts to avoid such a theory of adaptation by way of a theory of invention that truly allows for strong emergence. His notion of invention is radically creative because it recognizes the individual not as having possibilities in relation to a stable environment (in the way an Aristotelian substance has accidents) that might become actual, but as involved in ongoing transductive relations with its interior as well as with its milieu and the pre-individual. As such Simondon's theory combines a double movement reflective of this double relationship, which involves both the integration of the environment into the individual as well as the resolution of problems that the individual encounters by way of inventive resolutions at higher levels that can't be described as merely adjacent because it doesn't just involve an individual adapting *in relation to* an environment, but the individual-milieu-pre-individual *in* a process of invention.

It is precisely here that we can see the radical nature of Simondon's theory of individuation and the importance of the operation-structure relation that enables both an account of closure and regulation as well as a way to understand the radical openness of systems and that this entails the possibility of strong emergence or invention.

INFORMATION

Given the concern of molecular biology with studying genetics over the past half-century, it is perhaps surprising that Kauffman has relatively little to say about the subject. This is because he is predominantly interested in developing an account of life that precedes the mechanisms of genetic inheritance. Hence he investigates the possibility of life developing via the emergence of autocatalytic sets at a molecular level. By following this line of investigation Kauffman shares a similar methodology to Simondon in looking at how individuation emerges from the pre-individual level. It is perhaps also unsurprising that when discussing the concept of information, like Simondon, he concludes that there are two different operations being described that shouldn't be collapsed into the same concept.

We have seen that Simondon makes the distinction between primary and secondary information. Secondary information describes the classical cybernetic account of information (as developed by Shannon) relating to the transmission of a message between sender and receiver. However, primary information describes the actual transductive operation of individuation by which individuals (such as senders and receivers) actually come into being.

Like Simondon, Kauffman accepts that secondary information exists and plays various important roles but also that it is not primary enough to play a role in connecting matter and energy in the foundational process of morphogenesis. Thus, for Kauffman, Shannon misses that the receiver of information in the cybernetic sense must also be an agent. Thus he agrees with Simondon, who regards Shannon's notion of information as secondary and that can only come after the primary information that is individuation. Kauffman echoes this sentiment when he writes: 'The problem with applying Shannon's information theory itself to biology and the evolution of the biosphere is that we cannot make sense of the source, channel, or receiver' (Kauffman, 2010: 95).

Kauffman goes on to develop an account of a more primitive type of information that would be influential in ontogenesis. Developing an idea of microcode initially proposed by Schrödinger, he proposes a type of information that is coherent with the notion of morphogenesis already discussed and recognizes the importance of the topological. Thus he understands Schrödinger's microcode as a more fundamental kind of information, which is a

> highly heterogeneous set of *microconstraints* that are *partially causal* in the myriad organized events that are unleashed in the cell and organism in its propagating organization of processes. In this sense, information is nothing but the constraints themselves. This interpretation has the merit that it unifies information, matter and energy into one framework, for constraints are also boundary conditions. (Kauffman, 2010: 97)

What Kauffman develops here then is an account of information that recognizes the importance of topological conditions for directing work and thereby influencing the development of diversification in systems. As such Kauffman's account of primordial information resonates with Simondon's in that it also focuses on the resolution of disparities between energy and topology in his allagmatics as the science of the relations between operation and structure. The key is that for Simondon information refers to the *process* by which constraints come to be individuated and then modulate and organize further individuation, and not just to the constraints themselves.

AGENCY

Like Simondon, Kauffman is also aware of the paucity of an account of semantics in Shannon's account of information. Although Shannon's theory is useful for describing the transmission of a message, it fails to give an account of how the receiver obtains meaning from that message. This is the same criticism that we saw Simondon make in chapter 1 in that the problem is that this form of information fails to describe a true relation to that which is external to the receiver, that is, to its milieu. I shall develop Simondon's account of this in the next chapter with his theory of psychic individuation and the image.

It is worth noting here that Kauffman also develops just such an account of secondary information, although by way of Pierce, as 'the discrimination of a sign, say, of a local glucose gradient, and interpreting that sign by an action, say swimming up the glucose gradient' (Kauffman, 2010: 96). For Kauffman such a reaction by a bacterium is an example of agency and also constitutes an example of teleological causation. Meaning enters the picture because it is derived from agency. Kauffman gives the minimal definition of a molecular agent as

> a nonequilibrium self-reproducing system doing work cycles, to receive the information, discriminate it, and interpret and act on it. (96)

For Kauffman, then, it is via this operation of the discrimination of that which is external to the organism that meaning emerges, that then an *ought* can be derived from an *is*. As we will see in the next chapter, Simondon develops a somewhat more sophisticated account of meaning but nonetheless one that emerges at the same molecular level as Kauffman's.

NOTE

1. This metastable condition is what Kauffman calls *criticality* and denotes the same types of energetic condition that were described in chapter 2 when discussing systems that were close to falling out of phase with themselves.

Chapter Four

Psychic and Collective Individuation

In previous chapters I explained how in the physical and vital regimes individuation occurs by way of the resolution of problems via a process of disparation and integration resulting in the development of structures at higher levels. Simondon continues this genetic axiomatic in his description of the third regime of individuation, that of the psychic and collective. In this chapter I will describe Simondon's theorization of this regime by focusing on two of his texts: *L'individuation psychique et collective* (1989) and *Imagination et invention* (2008).

Each of these texts approaches the psychic from a slightly different perspective. *L'individuation psychique et collective* is of particular interest because of the concept of the transindividual, which Simondon develops there, but also for the ontogenetic description it provides of the development of the individual through emotion, perception and action. *Imagination et invention* is important because of the theory of the imagination, which it develops along with the related notion of invention.

AFFECT AND EMOTION

In his remarkable text *L'individuation psychique et collective* Simondon sets out the genetic development of affect and sensation, into emotion and perception, onwards to psychism and transindividuality as a progression through a series of problematics. This progression should not be understood as merely linear and straightforward. Although there is a progression, for example from affect to emotion or from individual to transindividual, this shouldn't be understood as merely a matter of accumulation. What Simondon describes is rather a number of problematics whose resolutions are never complete, but rather lead to further problems, which can also endanger previous achievements. In short, this is an exemplary application of an ontogenetic, transductive method that takes seriously the implications of metastability for the recurrent interchange between structure and operation.

Simondon describes emotion, perception and action as the three dimensions of the living animal. However, to understand the importance of these dimensions as well as their interrelation requires an ontogenetic account of their development. As always, Simondon insists on the need to account for genesis. It is for this reason that affect plays such an important role for Simondon, given its foundational role for the individuation of the living individual.

It is the affective that provides the subconscious grounding for psychism. As such, Simondon is fundamentally at odds with phenomenology's prioritizing of the intentionality of perception, such as proposed by Husserl. Perception, as it were, arrives later on the scene and requires its own account of grounding. To understand this it is important to understand that psychic individuation develops from an affective-emotional relation of the individual to *both* itself and the world.

In the previous chapter I described how the dynamic topology of an organism creates a resonance that is sustained throughout the interiority of that organism and establishes a metastable relationship between its interior and exterior. Even at the level of the simplest organism Simondon recognized the emergence of a rudimentary agency developed from the disparity between an organism's interior and exterior as well as the ability of the organism to police the permeability of its polarized membrane.

For Simondon, affect is that shift in polarization an organism undergoes due to disparities that occur within itself over time, due both to its inherent metastability as well as its relation to an exterior via sensation. The more sophisticated an organism's organization is, the more complex its chrono-topological relations will be.

> The affections are an orientation of a portion of the living being in relation to itself; they realize a polarization of a determined moment of life in relation to other moments; they coincide to being with itself across time, but not with the totality of itself and its states; an affective state is that which possesses a unity of integration to life, it is a temporal unity which is part of a whole, according to what one might call a gradient of becoming. (Simondon, 1989: 119)

As such affect needn't be conscious, in the sense of involving conscious mental activity, but nonetheless can be the cause for action. Kaufmann's example of a bacterium moving up a glucose gradient is a good example in that the affective awareness of the sugar in the environment creates a transductive relation between the sensation of the presence of sugar and the action of movement in response. Affect then is the fundamental way that an organism orients itself within its environment. It operates between two different orders of magnitude and is signified in each of them differently: from the perspective of the organism it is signified as a change in its internal resonance and from an environmental perspective it is signified by a change in the organism's behaviour.

It must be made clear, however, that affect does not describe a linear causal description by which a certain sensation thereby entails a certain mechanical effect from the organism without mediation. To adhere to such mechanism would precisely lose the importance of the temporal relation the individual has with itself and which is the ground for any emerging psychism.

Affects are polarizations that occur within the organism and by which the organism orientates itself both in relation to itself and to external influence. It does not entail the relation of a subject to an object but rather describes the organism's orientation to gradients of intensity and the disequilibrium this may produce for the organism's interiority. As such, the problematic of affect relates to how sensation challenges the temporal unity of the organism, thereby demanding the need for orientation and action. However, orientation is not based on the perception of discrete objects, affect develops prior to perception; rather the organism is fundamentally affective in its orientation and action in response to variations in environmental intensity and sensation.

The philosophical import of Simondon's ontogenetic account of the individuation of consciousness from the topological and chronological dimensions of the organism is significant and mirrors a fascination with ontology and biology that Simondon's doctoral supervisor, Merleau-Ponty, was also researching at the time. Both Merleau-Ponty and Simondon shared a concern with demonstrating that consciousness is immanent to the activity of nature. For Merleau-Ponty this was necessary to demonstrate how the gap between consciousness and object was not foundational. Although Simondon also shared this goal, more fundamentally he needed to demonstrate the establishment of the psychic regime from that of the operation of the vital regime of individuation.

Simondon shared Merleau-Ponty's fascination with biology for providing examples of the tropistic behaviour of organisms in response to sensations, as opposed to mere mechanical reflex. For Simondon it is the problematic the organism faces when experiencing conflicting tropisms or the disequilibrium of affects that leads to a disparity and the need of the creation of new unifying structures. It is emotion that marks the next ontogenetic step by coalescing disparate affects into more complex and fecund structures that enable the organism to orientate itself more meaningfully in its environment. Emotion is the modulation of the transductive process of affect, which structures the relation the organism has with itself as well as to what is outside it:

> The living being can be considered to be a node of information that is being transmitted inside itself—it is a system within a system, containing within itself a mediation between two different orders of magnitude. (Simondon, 1992: 306)

The internal resonance of an organism is the result of the organization of the parts of the organism into a unity and enables the signification of the difference between this unity and what is external to and exceeds it. However, this unity of the organism is never contained, it is always *both more and less* than a unity given the metastability internal to the organism that always has the potential for further individuation, but also due to its emergence from and continuing role in the broader individuation of the environment.

Over time, the internal resonance of the living being, as well as its relation to what is external to it, can also lead, via recursion, to further internal developments and external action, as well as transductive opportunities between them. Emotion is such a structuring within the organism that occurs as a resolution of the conflict of various affections. Emotion is the structuring and coalescing of heterogeneous affections that aid the unification of the polarization of the living being. As such it can be understood as a structuring of the subjective aspect of the living being that gives it a temporal coherence by which it can make sense of the world.

As we will see, however, it is incorrect to just understand the emotional structuring of affective being as a purely subjective affair occurring within the living being. Emotion is also structured in concert with the individuation of the collective. With Simondon it is always necessary to bear in mind that his is a theory of the psychosocial, that the psychic, especially at the level of emotion, is co-emergent with the social. I will explore this further in my discussion of the transindividual below, but first it is necessary to look at Simondon's theorization of perception, which describes how an affectivo-emotive organism transductively relates to the world.

PERCEPTION

Broadly speaking, affect can be described as the transductive process that brings into being the structure of the subject in that it concerns the operation of the modulation of information within the living being. In similarly broad terms, perception is that transductive process by which objects attain structure, that is, how the objective world attains structure for the organism. However, such general descriptions must be treated with caution given that both operations are intermingled in psychic development.

In *L'individuation psychique et collective* Simondon begins his discussion of perception by asking how it is that a subject 'seizes separate objects rather than a continuum of confused sensations, how it perceives objects with their individuality already given and consistent' (Simondon, 1989: 73). In developing his answer to this question he first considers three other candidate theories: associationism, Gestalt theory (which he calls the theory of the good form) and Shannon's information theory.

The theory of associationism holds that the ability to perceive objects rather than just an undifferentiated continuum of sensations comes about through the habitual use of analogy and the association of one experience with another in a learning process. As such it has some similarity with Hume's recourse to habit as a foundation for prediction. Just as Hume's use of the *synthetic a posteriori* is open to scepticism so is associationism. One can question how the initial perception upon which others are associated comes to be. If to perceive is always to engage in an analogical or comparative act then it is hard to see how perception can get off the ground. And indeed these are just the kinds of concerns that troubled Kant regarding Hume.

For Simondon associationism is also inadequate as it fails to give an account for how objects appear to have internal coherence rather than just remaining as 'an accumulation of elements'. Essentially, this is a repetition of his criticism of atomism but applied in the perceptual domain.

It is between the accounts of form that Gestalt theory and information theory provide that Simondon finds a tension in need of resolution. Where Gestalt theory gives an account Simondon describes as concerned with the 'quality of information', information theory's concern is with the 'quantity of information'. Simondon will situate his own account of perception as mediating the tension between these two theories via his theorization of information as *intensity*.

Gestalt theory is concerned with the quality of form due to it seizing form as a whole 'by virtue of a certain number of laws (e.g. the law of pregnancy, of good form)'. Simondon finds this approach questionable in that the wholes that Gestalt theory is concerned with are 'merely structural', that is, they do not necessarily refer to metastable and energetic wholes that are required by Simondon to form a system. Furthermore, and echoing his critique of hylemorphism, this focus neglects the genesis of form only to concentrate on the quality of the form as it is already perceived. This is not only a problem in accounting for past genesis but future ones as well:

> If the form was truly given and pre-determined, there would be no genesis, no plasticity, no uncertainty relative to the future of the physical system of an organism, or a perceptual field; but this is precisely not the case. There is a genesis of forms as there is a genesis of life. (Simondon, 1989: 74)

Semantics is also a problem for Gestalt theory, despite one of its laws for judging form being that of pregnancy, that is, whether a form is pregnant with meaning or not, it still fails to account for the genesis of this meaning as well as the forms themselves.

In contrast to the focus on the *quality* of information that interests the Gestaltist, Simondon describes the cybernetic approach to perception as focused on the *quantity* of information. As we saw in chapter 1, Shannon's cybernetic account of information is concerned with the technical problem of signal transfer. One issue with this account of information is that it fails to provide an explanation for how signals come to have meaning.

Another issue is that there isn't a necessary connection between the quantity of signals (however one wishes to quantify this) transmitted by a perception and how expressive or meaningful that perception is. Without this connection between form and quantity no solution is forthcoming regarding the problem initially posed as to why we see coherent objects rather than a continuum of sensations:

> The transmission of the image of a pile of sand or an irregular surface of granite rock demands the same quantity of signals as the transmission of the image of a well-aligned regiment or the columns of the Parthenon. (Simondon, 1989: 87)

Simondon's solution to this problem is to introduce another conceptualization of information that is applicable only to 'subjects directed by a vital dynamism' and that is information as intensity. To understand this it is necessary to take into account the 'whole subject' in a concrete situation in which the perceiving subject is equipped 'with tendencies, instincts, passions, and not the subject in the laboratory' (Simondon, 1989: 88). For Simondon the other theories of perception neglect the role of the polarized nature of the organism (its affective field), its metastability, as well as that perception is a system of individuation developed between subject and world the entirety of which 'constitutes the unity of perception'. This system involves the continuing and recurrent activity that attempts to resolve the tension between the individual that is more-than-itself (due to its internal resonance) and the metastable pre-individual that is its world and that exceeds it.

It is worth noting here that the subject as theorized by Simondon is a metastable system that individuates via a series of phase-shifts, thus when specifying the subject we are not describing something complete and wholly stable. It is a system always in operation and shifting depending on the problems and tensions that its individuation encounters.

At the stage we are describing here the subject for Simondon involves two processes of concurrent recursion that are in a problematic relation the ongoing resolution to which is found in the subject. These two processes are the recurrence of affect and of sensation, each of which develops transductively along a series of gradients into emotion and perception. This means that perception, far from concerning the distinguishing of stable pure forms or quantities of information, involves a state of tension involving heterogeneous factors.

Simondon's notion of the intensity of information aims to describe this tension and the process of its resolution. Thus although good form or the amount of information involved in a perception may be relevant, they don't necessarily make that perception meaningful for a subject:

> It is not necessarily the simplest and most geometric image which is the most expressive; it is also not necessarily the more detailed image, the more meticulously analysed in its details that makes the most sense for the perceiving subject. (Simondon, 1989: 88)

On the contrary it is quite possible that an image reduced in both the quality and quantity of information can have more intensity for a perceiver. Thus 'a slightly blurred photograph may have more value and intensity than the same photograph with perfect gradation respecting the value of every detail, or the geometrically centred photograph without deformation' (Simondon, 1989: 88).

This is because the intensity of information relates to the perception of the organism *qua* organism—that is, as vital, affective, emotional and world-orientated. Perception can't be distilled into a technical operation but requires theorizing for the role it plays regarding the organism and its relation to the world. Such an observation may seem mundane given how one's mood or even the influence of drugs can change the 'perceptual polarity' of how one perceives certain things but nonetheless it is something cybernetic information theory fails to take into account.

The question still remains regarding how it is that the subject comes to perceive the world as more than a confused continuum of sensations. In answer to this Simondon argues that as a transductive relation between subject and milieu it is also one of invention, in this case of form.

> Before perception, before the genesis of form which is precisely perception, the relation of incompatibility between the subject and the milieu exists just as potential. . . . Perception is not the seizure of a form, but the solution of a conflict, the discovery of compatibility, the *invention* of a form. (Simondon, 1989: 76)

Perception is then the resolution of a problematic. The invention involved in perception is the organization of intensive thresholds and disparities to establish the coherency of the object. Such intensity, as we have seen, involves the subject's orientation to the world as well as the retention of its perceptual history. In a similar way to that by which recurrent sensations can develop into perceptual activity, past perceptions can have a conditioning effect on present perception. This brings Simondon close to a transcendental aesthetic such as found in Kant, although one which is sharply distinguished from Kant's in that Simondon demands a genetic account, which Kant would condemn as a mere 'physiological derivation', answering the question *quid facti* (CPR A87/B119), of any such development and in doing so also intimately connects the subject, body, emotion, action and world. I will more closely consider this development later when I look at Simondon's theory of the image that concerns itself with the genetic development of the organization of the imagination. However, with the theorization of intensive information it is clear that Simondon understands perceptual activity as a kind of vital individuation that is not reducible to the cybernetic notion of information.

Simondon also provides the example of binocular vision that is useful for understanding how the organization of intensive thresholds and disparities can result in a higher systematic. In this example the disparation in need of resolution is the slight difference between the left and right retinal images. In binocular vision this difference isn't discernable because both images are resolved into a higher systematic that is signified as depth. As such signification is understood as the result of a resolution of disparity between two signals. If there is no disparation between signals there will be no information or signification as there is nothing being resolved into a higher functioning. Conversely, if the disparity is too great it is more likely that a resolution isn't possible and 'information suddenly becomes zero when the operation by which the disparation is assumed as disparation can't be performed' (Simondon, 2013: 222).

THE PSYCHIC

In the situation of the purely vital, which Simondon calls 'simple life' (Simondon, 2013: 165), affectivity maintains a unified coherence between perception and action enabling the orientation of the living being. However, sometimes the unified polarity of the organism is challenged by sensations that lead it to develop more highly unified and structured emotions. This occurs in situations in which 'all the problems of the living being can't be resolved by simply regulating transductive affectivity' (Simondon, 2013: 165) within the interior of the living individual. In these situations affectivity cannot maintain its role as the regulator of vital individuation 'which is life itself', but instead,

> it creates instead of resolving problems, and leaves unresolved those of perceptual-active functions. The entry into psychic existence manifests itself essentially as the appearance of a new problematic, much higher, more difficult, which cannot receive any real solution in the interior of the living being

> properly speaking, conceived within its limits as an individuated being. (Simondon, 2013: 165)

In this situation 'affectivity is overflowing' the regulation of vital individuation occurring in the interior of the living individual and instead is active in establishing new problems for perception and action, whose resolution requires a 'superior rearrangement of vital functions' in a new 'more primitive individuation' that occurs following a new 'dive into the pre-individual reality'. This new individuation is that of the psychic, which Simondon describes as a 'precocious expansion of vital individuation' (Simondon, 2013: 166).

As well as drawing from the pre-individual this new individuation of the individual, which can be understood as a phase-shift, that is, a quantum jump to the psychic regime, also draws from its relation to the collective. This relation of the psychic to the collective cannot be stressed enough:

> The psychological world should be named the transindividual universe rather than psychological world, because it has no independent existence; for example, culture is not a reality that exists for itself; it exists only in so far as monuments and cultural testimonies are reactualized by individuals and understood by them as carriers of meanings. (Simondon, 2013: 272)

Psychical activity must not be understood as the emergence of a new substance, such as described in Cartesian dualism, but as the development of a new type of vital functionality that emerges in response to problematic situations organisms finds themselves in. It is a dilation of vital individuation. As such the structuration of a psychic domain should be understood as a new kind of vital relation to a milieu developing out of vital functionality.

This structuration isn't the realization of a potential limited to only certain vital beings, such as being restricted only to humans. Simondon is clear that psychism is not an essential characteristic of certain species but rather that many animals might 'find themselves in psychical situations, only those situations which lead to acts of thinking are less frequent in animals' (Simondon, 1989: 152fn). Additionally, this non-essentialist perspective regarding psychism also means that it cannot be utilized to 'found an anthropology' because psychical ability is not an exclusively human trait. Some animals are capable of psychical activity and the extent of psychical activity varies not just from animal to animal but also from person to person. The psychical emerges when an organism faces a problem that disrupts its seamless affective somatic relation with its milieu:

> Affectivity indicates and comprises the relation between the individualized being and pre-individual reality: it is therefore to a certain extent heterogeneous in relation to the individualized reality, and appears to bring it something

> from the exterior, indicating that it is not a complete whole and closed reality. (Simondon, 1989: 108)

This disruption is also a result of a neotenic 'deceleration of the living', 'which keeps it in a metastable and tense state, rich in potential' (Simondon, 2013: 165), such that the psychic can emerge as a response to this intervention in vital operation. The individuation of the psychic is also described as a process of *individualization*. This denotes psychic individuation as the further individuation of an individuating living individual. That is to say that individualization is a phase that does not replace vital individuation but operates simultaneously alongside it. However, it must be remembered that individualization is also an ongoing process and is never complete. This further individuation can be understood as differing from that of the purely vital regime in that it results in the structuration of the individual into *both* somatic and psychic domains. As such, it is through this process of individualization that the subject emerges, which is marked by the individual developing a personality.

To understand what personality means for Simondon requires being clear regarding the difference between individuation and individualization *apropos* the individuation of the psychic. Individualization marks the attainment of a new spatio-temporal systematic that resolves the problem of the 'overflowing' of affect for the vital individual at a lower level. This problem is resolved by the organization of the new operational structure of the psyche. This production is also the production of signification—that is to say that the development of the psyche is meaningful as the resolution of a problem or disparity through attaining the production of a new axiomatic of functioning. As such Simondon connects the process of individualization with the distinction between signals and signification in which signals are that which pass between individuals but 'the individual is that by which and in what appear significations'. The individualization that is the psyche both reaches to the exterior and interior of the living being and as such echoes the individuation of the vital being but at a new level, discovering solutions that are posed of the living from a 'plurality of signals' and which 'increases the intelligibility for it its relation to the world' (Simondon, 2013: 257).

The individual being can now be understood to be composed concurrently of 'the individual being qua individuated being' *and* 'the individual qua individualised being' (Simondon, 2013: 258) and it is the personality that Simondon asserts is 'all that attaches' these two individuations of the individual.

We must be careful not to separate the operation of these two domains, the somatic and psychic, too rigidly, for it should be clear that the operation of one is dependent on the other. As such Simondon provides a solution to his supervisor Merleau-Ponty's question of how 'a given fact of the objective order' (he uses the example of a cerebral lesion) can disturb the 'given relation with the world', that is, consciousness (Merleau-Ponty, 1968: 200). Simondon also briefly applies this schematic to sexuality where he contends that a sexual relation is to be found at the level of individuation whereas at the level of individualization the relation 'is of the type that is afforded by the contingent events of everyday life' (Simondon, 2013: 258). At the level of the personality a relationship would be 'that integrates in a unique situation sexuality and individual event history' (2013: 258), the implication being that sexuality is both given somatically but also by the contingent aspects of our personal history, which may colour our preferences and actions. A truly human relationship would combine these two aspects of the whole personality. In contemporary discourse it's possible to read this as the tension between biological sex and gender, in which case Simondon would clearly be proposing some measure of physical determination in sexuality and certainly rejecting a *wholly* discursive account.[1]

Before going further it is necessary to address the relation that holds between the individual and the collective. The development of subjectivity occurs for the individual, with and through the collective in a process of orientation within its milieu via signification; as such the psychic for Simondon is always actually psychosocial. The ability of the individuation of the psychic personality to maintain itself depends on the functional support of the cultural milieu within which it is situated. This is not to say that the psychic personality is over-determined by the collective. Although the collective relation is key for helping the individual overcome the affective problems it faces it is also the source of fresh problems:

> However, the psychological world exists to the extent that each individual finds before him a series of mental and behavioural schema already incorporated into a culture, and which incites the posing of their particular problems according to a normativity already elaborated by other individuals. (Simondon, 2013: 272–73)

This relation to an exterior that sustains can also lead to the individual becoming decentred by what exceeds it. In that situation a new problematic arises regarding the psychic individual's ability to resolve itself in the collective.

THE TRANSINDIVIDUAL

It is because of the need for this further resolution that Simondon claims that the psychic should also be understood as the 'nascent' transindividual, that is, that individualization is a step towards the transindividual. This doesn't, however, mean that individualization, the coming into existence of the psychic subject, is also the emergence of the transindividual. Simondon distinguishes between two modes of relation that occur *between* psychic individuals: the inter-individual and the transindividual, thus making it clear that the transindividual relation indicates a specific accomplishment that is required for the resolution of a new problem that the psychic individual encounters, which is the disparity it finds between itself and that which exceeds it, both in respect to its environment and to other psychic individuals.

Simondon names this situation of incompatibly the 'problematic of embodied immanence [*problématique d'immanence incorporée*]' (Simondon, 2013: 294). In attempting to resolve this affective problem within itself the individual undergoes a state of *anxiety*. That is, Simondon claims, as a solitary subject it attempts to subsume within it all that is exterior to it. Such an attempt is necessarily doomed to failure.[2]

With the impossibility of the subject finding a resolution for this problematic within itself it becomes clear that it is through individuation with the collective that this problem is to be resolved: 'The collective taken as axiomatic in resolving the psychic problem corresponds to the notion of the transindividual' (Simondon, 1989: 22).

The difference between the inter-individual and the transindividual relationship is that the former is unable to resolve this problem because it is a relation that 'goes from one individual to another' but 'it does not penetrate individuals' (Simondon, 2013: 294). Instead the inter-individual relationship is understood as simple in that in it individuals relate to one another via the representations they have of one another as if they are fully constituted individuals. As such, there is the passing of signals without signification. The inter-individual relationship is one that is based on established norms and functional relationships. That is to say that it is the kind of relationship that individuals have with one another when they accept and understand one another in terms of the established shared values and behaviours of their culture. It is a relation that can 'remain a simple relation and avoid reflexivity' (Simondon, 2013: 273). Simondon also includes in this description the relations often involved in labour due to these relations being the result of a functional organization imposed hylemorphically on a workforce, and by doing so is deliberately undercutting the importance Marx gives to labour relations. However, as Simondon states, 'everything is not given in culture' (Simondon, 2013: 273) and even in the most normative cultural circumstances an individual will still eventually come to question itself. It is through such reflexive questioning that the individual becomes decentred and is thus in a position to individuate beyond inter-individual relations towards the transindividual. It is for this reason that Simondon states that the 'inter-individual relationship can hide transindividual relations' (Simondon, 2013: 273).

In the transindividual relation, the individuals are said to form a system with one another in which there is an ongoing resonance between them:

> Transindividual action is what makes individuals exist together as the elements of a system including potentials and metastability, anticipation and tension, then the discovery of a structure and a functional organization that integrates and resolves this problematic of embodied immanence. (Simondon, 2013: 294)

The transindividual is a mode of relation that is established between psychic individuals in that in doing so there occurs a *co-individuation* and coincidence of those in relation to a shared pre-individual milieu. The functionality of the group emerges from this relation and is not imposed upon it, as in the situation of labour. This co-individuation, which distinguishes the transindividual, entails the individuals of the collective forming a system involving the simultaneity of their actions as well as their sharing an emotional charge. The establishment of the transindividual level is also described as an auto-constitution in a manner similar to that of the emergence of the psychic from the vital, in that the psychic is an expansion of the vital individuation that exists concurrent to it, never wholly separated from it:

> The relationship between the individual and the transindividual is defined as *exceeding the whole individual in prolonging it*; the transindividual is not exterior to the individual and yet is detached to a certain extent from the individual. (Simondon, 2013: 274)

The transindividual relation develops in a double movement by connecting the interior of the individual to the exterior ('interiorize the exterior'), while also connecting the external to the individual's interior ('exteriorize the interior'). This double connection is established across the group through the co-individuation of emotion and action. This becoming social of the psychic is thus clearly demarcated as the *psychosocial* insofar as what is being described is a relation that is constituted between two poles in the sense that each (psychological/social) is just a description of the same relation (the social relation) from a different perspective. For this reason Simondon, continuing his habit of getting beyond dualisms, and also as part of his reformation of the social sciences via axiomization, questions the undertaking of psychology and sociology as separate disciplines. For him each discipline's area of study is situated at the extreme pole of the same social relation, and as such each limits itself from understanding the entirety of what they claim to study. What the term *psychosocial* indicates is the presence of an obscure zone of operation that occurs beyond and between the disciplinary boundaries of psychology and sociology, the recognition of which would make clear the need for their ontogenetic unification. Thus psychology approaches its subject purely from the perspective of the interiority of the subject, thereby neglecting the reality of exterior relation, that it is both a *mediation* of interiority and exteriority. Sociology, on the other hand, substantializes the determining nature of the social on the psychic reality of individuals. Both subject areas over-determine the importance of the pole on which they concentrate to the detriment of gaining an understanding of the relation that is constitutive of the phenomenon they study, which is the social relation of transindividuality. Instead of understanding the psychological and social in a transindividual situation as co-constitutive and in constant negotiation, the relationship is viewed as atomistic or hylemorphic, depending on the theory.

Although Simondon's critique of sociology and psychology may seem rather general given some more recent developments in these fields, including the development of the psychosocial as a field of study in its own right,[3] it does indicate that with his goal of encyclopaedism he presaged today's growing awareness of the interdisciplinary. The interconnectedness of all things has today become something of a cliché and in its holistic form as unhelpful for thought as reductionism. Simondon's attempt to think through how things interconnect, that is, what the nature of relation is, without resorting to reductionism or holism, is one of his great achievements.

From an ontogenetic perspective, the relation of the individual to the collective also needs to be understood temporally. That is, an individual's relation to both past and future must be understood through the structuring relation it has with its social group. In order to describe this Simondon develops a schema that maps the reality of an individual's social relation. Initially this schema consists of two groups, the *group-of-interiority* and the *group-of-exteriority*.

To participate in the *group-of-interiority* requires that an individual's past and future resonate with others in the collective. In order to do this the individual must have a shared heritage with the group and also share in the direction for future collective action. The delineation of what constitutes a suitable past or heritage for a group is diverse in that it might be composed of a shared set of traditions, laws or politics or even around objects of shared interest such as religious or academic texts. The point is that via this heritage the collective has a shared past as well as a horizon for shared future action. It is not to say that an individual stands alongside another but rather that all individuals in the group are in a state of resonance brought about by their being penetrated by social relations that constitute them as a subjects in that group; for example, this can be experienced by the subject through belief:

> Belief, as a mode of group membership, defines the expansion of the personality to the limits of the in-group, such a group in effect can be characterized by the community of implicit and explicit beliefs among all members of the group. (Simondon, 1989: 177)

We can now return to the problem of anxiety that we saw was the problem for 'embodied immanence' in that the individual had no way to resolve the emotion of going beyond itself through action. A transindividual relation that is maintained through a *group-of-interiority* enables a resolution to this anxiety in that there becomes available to the individual a course of action as part of a group that enables the individual to extend itself. A transindividual relation can thus be understood from different perspectives such that an 'action is collective individuation seized on the side of the collective, in its relational aspect, whereas the emotion is the same collective individuation seized in the individual being as it participates in this individuation' (Simondon, 1989: 107). So although emotion has its roots in the affectivity of the individual, it is also constituted via that individual's relation with the social and 'neither the pure individual nor the pure social can account for emotion which is the individuation of pre-individual realities at the level of the collective instituted by this individuation' (Simondon, 2013: 306).

Additionally, it is now clearer how affect and action require a relation to the psychosocial in order to enable the individual to extend itself through emotion in the collective, thereby resolving its anxiety:

> The essential instant of emotion is collective individuation; after that instant or before that instant, one cannot discover the true and complete emotion. Emotive latency, inadequacy of the subject to itself, incompatibility of its charge of nature and of its individuated reality, indicates to the subject that it is more than individuated being, and it contains within it the energy for subsequent individuation; but this subsequent individuation can not be achieved in the being of the subject; it can be done only through this being of the subject and through other beings, as a transindividual collective. (Simondon, 2013: 306)

In contrast to the *group-of-interiority* is the *group-of-exteriority*. Unlike the social relation a subject has in a *group-of-interiority* there is not an isomorphism between the past and future of the individual with the past and future of the collective in the social relation of the *group-of-exteriority*. Instead the individual does not experience the collective but rather the social. Although sometimes inconsistent in use it is worth drawing out Simondon's use of terminology here. A subject, for Simondon, is an individual that has been able to make use of its pre-individual charge through an operation of transindividual individuation. Thus he writes:

> The entry into the collective must be conceived as a supplementary individuation, utilizing a charge of pre-individual nature which is borne by living beings. (Simondon, 1989: 215)

The subject is then an individual that has individuated with a collective thus resolving the problem of anxiety. However, an individual who has not made use of the pre-individual charge to resolve the affectivo-emotional problem through a transindividual relation is not seen to be a subject but instead finds itself unable to co-individuate. As such Simondon describes such an individual as relating to a *group-of-exteriority*. Such an individual cannot share in the heritage or past that the collective does, the past appears to it as 'a set of isolated points' that can't be formed into a system. Similarly, its relationship to the future, instead of being open to a transductive collective relation of invention, is one in which the future is so set out as to condition the present. Due to an inability to pool its potential with others its ability to change the future or expand itself is restricted.

What Simondon is describing with these two groups is not actual social relations but the mode by which social relations develop and are experienced. Thus the *group-of-interiority* describes the relation of those who are firmly embedded in the collective. Conversely the *group-of-exteriority* describes the quality of those who are less integrated into a collective. As such these two groups mark the extreme poles of the transindividual relation, with the subject/collective at one end and the individual/social at the other. The social must not be seen in a substantial sense standing in opposition to the individual but rather as the outer limits of a 'zone of participation' (Simondon, 1989: 179) in which the individual participates, which has the *group-of-interiority* at its core. The *group-of-interiority* marks the 'social body of the subject' from which it orientates itself and in which it invests itself.

As well as it being possible for an individual to shift its relation on this gradient between the two groups over time, it is also likely that the nature of these groups will also change. For example, the *group of interiority* could narrow under the influence of a fascistic influence or it could broaden and become more inclusive. One of the powerful features of Simondon's schema is that not only does it enable a direct connection between the psychological and the social but it is also applicable to the theorization of collectivities of all size. There is also no reason why it can't be applied to the simultaneous co-individuation of multiple collectives.

Thus far we hope to have demonstrated that Simondon, in extending his transductive axiomatic to the psychosocial domain, has enabled an approach to the psychosocial that falls neither into the trap of reductionism or holism while also developing a framework in which to situate that which we find in both sociology and psychology. However, what we have yet to interrogate is what constitutes the pre-individual charge of nature by which the individual is able to go beyond itself in this schema and how it could be useful in our broader enquiry into Simondon's socio-cybernetics. As we have seen in relation to both physical and vital individuation, the pre-individual has been described in terms of energetics and disparity. It is less clear in relation to psychic and collective individuation what the pre-individual references. A passage from near the end of *L'individuation psychique et collective* does little to illuminate:

> This charge of individuated reality conceals a power of individuation which, in the subject alone, cannot succeed, through poverty of being, through isolation, due to a lack of an overall system. Together with others, the subject can be correlatively theatre and agent of a second individuation that is the birth of the collective transindividual and relates the subject to other subjects. The collective is not nature, but it presupposes the prior existence of nature attached to the subjects between which the collective establishes itself by recovering them. It is not really as individuals that beings are attached to one another in the collective, but as subjects, that is to say, as beings that contain pre-individuality. This doctrine seeks to consider individuation as a phase of being. This phase, also, might not exhaust the possibilities of pre-individual being, so that a first individuation gives rise to beings which still carry with them virtualities, and potentials; too low in each being, these potentials, joined together, may operate a second individuation which is of the collective, attaching individuals to each other by the pre-individual they retain and carry. (Simondon, 1989: 204)

In this passage Simondon is clear that the individual does not have enough pre-individuality within itself to undertake the further individuation of transindividuation by itself. What we have seen is that for the transindividual relation to establish itself what is required is a concurrence of action and emotion by which exteriority and interiority come into relation. Such a relation builds on the pre-individual present in the individual but also requires a supplement from outside the individual associated with action. It is because of this that the act of invention becomes important for the theory.

IMAGINATION AND INVENTION

In his course on general psychology from 1965 to 1966, which was later published as *Imagination et invention* (2008), Simondon develops a theory of the imagination as a genetic development of psychic activity from the vital. As discussed above Simondon was critical of psychology as a discipline due to its narrow focus on interiority. This criticality is maintained in this course in that his theory is directly opposed to those who maintain the centrality of the subject and representing consciousness.

One target for criticism is Sartre's theory in *L'imaginaire* (1940) of the transparent consciousness that produces representations in a form-giving hylemorphic manner. For Simondon, this understanding of imagination as a faculty of the subject gets the description back-to-front and is in error because it excludes 'the hypothesis of a primitive exteriority of images with respect to the subject' (Simondon, 2008: 7).

Although Simondon is sometimes described as a phenomenologist, this attribution is not entirely accurate. He is less interested in describing the relation of phenomenal consciousness to the world than describing the genetic process by which vital organization gives rise to, and is continued in a different mode by, the psychic activity upon which phenomenological consciousness is constructed. Thus phenomenology, including Sartre's version, lacks a suitable account of the ontogenesis of consciousness.

What Simondon proposes, regarding organisms with central nervous systems, is the reality of a pre-conscious (or at least subconscious) 'fund of images', which underpin and make possible the forming of representations. That is to say that images arise within the living being prior to and independently of consciousness, intentionality and perception. To understand this we need to describe the genetic theory of the cycle of images that Simondon proposes. The cycle has three phases: the first involves the development of an image, which is an 'embryo of activity and perception'; the second involves the development of a perceptual world along with various 'schemas' of response to it; the third involves a systemization that results in the development of a mental model of the world. Each phase of this image cycle can be 'related to a dominant activity or function'; these are anticipation, experience (perception) and memory.

Simondon, like Merleau-Ponty, made use of Gesell's work regarding the ontogeny of behaviour[4] and how it develops alongside the growth of the organism to inform his account of the first phase of the image. In this work, as well as in that of other biologists and ethologists (such as Coghill and Jennings), Simondon discerns the emergence of what he calls a 'motor-image of anticipation':

> But in fact, Gesell's studies have shown that the ontogeny of behaviour is similar to growth: not only is it made according to the principles of polarity, orientation, according to gradients, and not evenly like a balloon that is inflated, but it also occurs according to successive cycles separated by de-differentiations preparing new structuration. Each step results at the end of the cycle to a defined behaviour, which might stand on its own unless it was only a moment of a larger genesis; temporarily abandoned, it will be reincorporated in its essential lines to the definitive more complex synthetic 'pattern'. It is this existence of the essential lines of behaviour that can be considered as furnishing the content of the motor images of anticipation of the conducted. (Simondon, 2008: 38)

In the first phase of the cycle these anticipatory motor-images develop endogenously in the organism independently from one another in a 'free state' in a manner analogous to the growth of organs. As such, Simondon clearly indicates that this phase is biological or vital.

Although these images may be independent from one another at this stage of the cycle, they all develop in relation to an associated milieu. Simondon conceives of images at this anticipatory stage (that is, prior to experience) as activity that has been put in reserve, that is, as the developed behaviour of the organism it has acquired as it has grown; he describes such reserved behaviours as virtual. What Simondon has in mind here is the spontaneous development of patterns of behaviours (which are often described as instinctual) such as those related to typical situations like feeding, evading predators and mating.

One important consequence of this theory is the primacy of motoricity, that is, that an organism's motor abilities are developed prior to (or at the very least contemporaneous with) the development of perception. This priority holds from simple organisms to the more complex; from the unicellular organism's development of a polarity as an index of orientation to the human whose motoricity begins developing during its time as an embryo.

> To say that the motor precedes the senses, is to assert that the stimulus-response pattern is not absolutely the first, and that it refers to a situation of actual relation between organism and environment which has already been prepared by an activity of the body during its growth. Research by Jennings on the simplest organisms show that reactions (performed in the presence of an object) are preceded by motor spontaneities existing before receipt of signals characteristic of an object. (Simondon, 2008: 29)

The priority Simondon gives to motoricity not only places him outside the remit of phenomenology (at least as that characterized by Husserl) but also places him in a radically non-Kantian position regarding the question of the possible conditions for both perception and knowledge. As Mark Hansen observes,

> By defending the autonomy of the sensory, Simondon is able to correlate the image with motoricity *prior* to the advent of perception and to maintain its independence from object perception. (Hansen, 2011: 110)

This independence of the image is another radical difference in Simondon's theory. Unlike in representational theories of the subject in which the image is created by the subject, Simondon argues that the image actually has a relatively independent existence from the subject, to such an extent in fact that he describes the image as like a parasite; that is to say that although it is reliant for its existence on the organism the image develops within it independently.

Simondon describes the second phase of the cycle of the image as that of experience. In this stage the previously independent anticipatory images organize into groups according to the reception of information from the milieu.

It is at this stage that the images are described as forming a 'world' for the organism and attain a functionality of both stabilizing the relation between the organism and milieu as well as developing schemata of response to the world. It is clear that such an understanding of world for Simondon is close to how Merleau-Ponty utilizes von Uexküll's notion of *Umwelt* in developing a philosophy of immanence; that is, the *Umwelt* as the relation of an organism's activity with a milieu.

In this second phase the organism moves beyond merely anticipatory, 'instinctive' reactions and obtains 'the perception of its present state, with appreciation of variations and differences. . . . The image is used here as an instrument of adaptation to the object, it assumes that there is an object, and not merely a situation' (Simondon, 2008: 22).

Perception can then be understood as a function that emerges from this secondary phase of the image and also as the development of a psychology in the form of a mental analogue of a primary relationship with the milieu that enables the organism to move beyond just instinctive responses and organize responses to objects (planning/motivation). Perception is therefore never direct but relies on this fund of images, which are organized relative to experience of the object in a recursive fashion in what Simondon calls an apprenticeship. Such an apprenticeship involves the development of schemata to help organize responses to the milieu via experience.

The third phase of the cycle is that of a systemization of images, built on this apprenticeship, utilizing an affective-emotional resonance that

> achieves the organization of images in a systematic mode of links, evocations and communications; it creates a real mental world where regions, domains, qualitative key points by which the subject possesses an analogue of the external milieu, having also its constraints, its topology, its complex modes of access. (Simondon, 2008: 19)

This resonance is the development of the function of memory for Simondon and enables the 'ability to relive situations from the evocation of the image'. Additionally, this phase also involves a systemization of the images and the achievement of a 'real mental world'. It is here that Simondon comes closest to a kind of Kantianism in that he discusses the formal role that such a systematized image can have for action and perception.[5]

> Like anticipation the *a priori* image appears as a form of motor intuition, a scheme of projection from an active center of spontaneity and radiating towards the plurality of situations or objects. (Simondon, 2008: 22)

In chapter 1 I discussed Simondon's concept of technical mentality; that is, how schemata of thought have been developed through analogy with technical operation. Simondon cites Cartesian rationalism and cybernetics as examples. Such schemata are examples of the kind of systematic organization of images that can facilitate a formal cognitive relation to the world. It should be apparent that the development of such *a priori* images occurs ontogenetically across the lifetime of the organism and is clearly not transcendental in Kant's sense.

It is at this point that Simondon introduces the notion of invention as a fourth and final phase of the image cycle. As we have already described, the image is to be understood as a process involving formalization through a cycle of phases. Additionally, each phase of this cycle is related to a function (anticipation, perception, memory). The third stage consists in the systemization of symbolic images. However, any systemization can become problematic if it becomes saturated in such a way that it is unable to integrate new information. Such a situation can result in the dephasing of the current systemization resulting in

> a new state of free images allowing the recommencement of a genesis: the invention would be a renaissance of the cycle of the images, allowing an approach to the milieu with new anticipations from which emerge adaptations that were not possible with primitive anticipations and a new internal and symbolic systematization. In other words, the invention makes a change of level, and marks the end of one cycle and the beginning of a new cycle, each cycle consisting of three phases: anticipation, experience, systematization. (Simondon, 2008: 19)

Thus invention can be understood as a transductive operation in which the symbolic organization is dephased due to a disparity within the organization of a system of activity. This dephasing enables the recovery of a sense of anticipation regarding the necessity to invent afresh the imaginative organization of a world. Furthermore a true invention is one that involves the invention of an object, a process that Simondon understands as the making exterior of the process of the image (material images). That is not that the invented object was first a representation to a consciousness that has then been realized externally but that the exteriorization is a part of the process of the development of the image itself and in turn helps in the ongoing systemization of the image through its involvement in structuring the relation of organism with milieu. The image then is always the ongoing support of the operation of the organism-milieu dyad. As such we must also consider it to be a support for the transindividual:

> In effect, the image, as intermediate reality between the concrete and abstract, between self and world, is not only mental: it materializes, becomes institution, product, wealth, and is diffused as much through commercial networks as through the 'mass media' disseminating information. Its intermediate character, a fact of consciousness but also object, gives it an intense capacity for propagation; images permeate civilizations and charge them with their pow-

> er. . . . The circular causality that runs from the mental to the objective real through social processes of cumulative causation also runs from the objective real to the mental. Every image is susceptible to incorporation in a process of materialising or idealising recurrence; deposited in fashion, art, monuments, technical objects, the image becomes a source of complex perceptions awakening movement, cognitive representation, affections and emotions. Almost all objects produced by man are in some measure object-images, they are carriers of latent meanings, not just cognitive but also conative and affectivo-emotional. Objects-images are almost organisms, or at the very least germs capable of revitalizing and developing in the subject. (Simondon, 2008: 13)

In this remarkable passage Simondon connects within a recurrent causal relationship the gradations of psychical activity from sensation to ideation and transindividuality to the world of artefacts described as object-images. The reciprocal flow of causality between ideation and the 'objective real' describes a truly novel way of thinking the psychic, social and technology as systemically modulating and organizing one another.

As we saw in the passage from *L'individuation psychique et collective* quoted at the end of the previous section, the subject lacked enough resources in itself to make the leap to the transindividual but required the establishment of a relation to the collective. The transindividual relation is described as the concurrence of action and emotion. What Simondon's *Imagination et invention* suggests is that the act of invention provides objective supplements that can help initiate transindividual individuation in the form of objects, symbols and technologies.

CONCLUSION

The two accounts of psychic individuation described in this chapter, first in Simondon's main thesis and the second in his course *Imagination et invention*, differ in the perspectives and core concepts utilized. Where the account in the thesis is focused on the ontogenetic individuation of the psychosocial structured by transductive phases from the pre-individual onwards, the account in *Imagination et invention* concentrates on the ontogenetic development of the imagination as a function of the organism.

The differentiation between individuation and individualization in *L'individuation* provides Simondon with the means to also account for the difference between the transcendental and empirical. Understood as individuation, the transcendental indicates the shared and universal process of individuation that all beings have undergone and that underpins and guarantees the possibility for shared knowledge. Given that all individuation begins from the pre-individual and that thought is itself an operation of individuation that occurs within an individuating subject, there is a sense in which the universal rooting of all individuation provides a transcendental *a priori*. This is not to be understood in the manner of a Kantian *a priori*; it is not to claim that there is a stable *a priori* transcendental structure that is shared, thus normalizing and stabilizing knowledge. What individuation provides is the conditions of possibility for what knowledge can be, as knowledge itself is an individuation:

> If knowledge finds the lines for interpreting the world as stable laws, it is not because there are *a priori* forms of sensibility in the subject whose coherence with the raw data from the world through sensation would be inexplicable; it is because being as subject and being as object result from the same primitive reality, and that the thought that currently appears to establish an inexplicable relation between the object and the subject in fact only prolongs this initial individuation; the *conditions of possibility* of knowledge are in fact the *causes of existence* of individuated being. (Simondon, 2013: 257)

Additionally, transcendence cannot describe the externality of a form of superior being, whether that is understood as the pre-individual or transindividual. To think the individual as somehow separate to these is to miss the process of the individual's constitution within, and as part of, individuation; it is to misunderstand what the individual is and make the mistake of defining it through terms specified after its individuation has occurred. As such it is to repeat the mistake of hylemorphism again.

The empirical perspective is that of the subject from the perspective of individualization. If the transcendental as individuation guarantees the ground for a shared knowledge then the empirical is the perspective of the subject, which is reflexive regarding the problems it faces due to its relation to individuation. Its experience is, to a certain extent, singular, although, as we see from the account of the ontogenetic development of the imagination, is also structured in the collective.

The subject experiences itself as confronted by transcendental problems and has to choose how to solve these problems. One method for doing this is the invention of schemas of understanding by which to address them. This method involves the development of the imagination through the concretization of empirical knowledge regarding the transcendental, while also simultaneously being a new problem in the transcendental. It is in this way that we can also understand the cycle of the image as the simultaneous solution and problematizing of transcendence.

I have now completed my description of the main aspects of Simondon's ontology and associated epistemology. Already we have caught some glimpses of how technology can play a part in processes of individuation. For example, it offers resources, along with other forms of symbolization, to enable transindividuation. It is also a key component in the cycle of the imagination enabling the production of schemas for understanding the world.

The breadth of Simondon's ontology means that any account of technology must take into account the relations it is productive of across the broad range of individuations already discussed. Thus as object-image, it needs to be understood in its causal relation to all levels of the gradation of the psychic. There is already some work being undertaken to implement this Simondonian insight, especially for understanding contemporary digital technologies in relation aspects of the psychosocial just outlined. I will attempt to address some of these in the coming chapters.

For example, the anticipatory and motor dimension indicates that our engagement with a milieu must involve aspects that are pre-phenomenal. This is an area investigated in the recent work of Mark Hansen who, in attempting to theorize contemporary digital technologies, considers the 'mental image as microtemporal pattern of cognitive activity' (Hansen, 2011: 87), that is, at a level prior to that of discursive determinacy and phenomenology, for example.

Memory is also another important aspect of this relation and is something Bernard Stiegler theorizes in his work that views technology primarily as mnemo-technics. There is certainly much to commend this perspective although we will not draw the same conclusions as Stiegler regarding contemporary technology use if we try it understand it as involving a broader relation to the social than just at the level of memory, as the cycle of images suggests we must.

It is perhaps now clear that the scope for such an investigation of media technology is going to be extremely broad and complicated. But before I can attempt this I must first address Simondon's own perspective on technology, containing as it does another key factor that must be incorporated into my analysis, which is the reality of technical evolution as a causal factor itself.

NOTES

1. In her work Elizabeth Grosz points out that some recent feminist work has lost touch with the 'real' and 'rendered impossible the notion of a pre- or non-representational real, seeing in biology only fixation and resistance to change, and regarding what is creative as what is consciously created by human intentionality' (Grosz, 2012: 52). Grosz utilizes Simondon to get beyond representation and identity politics based on stable identities to instead question the giveness of individuals (be they 'biological, social or collective') in order to enable new ways of understanding and addressing the nature of what constitutes individuals and collectivities. Thus Grosz seems to accept Simondon's account of sexuality as the formation of a new problem involving both individuation and individualization: 'Cultural "gender" is the transcription, at another level, of the tensions and sources of upheaval posed by sexual selection at the level of animal or vital existence. In this sense, it functions in different terms from all other forms of social collectives; it is a problem, an irresolvable tension of animal life that is animated and transformed, negotiated, in socially variable ways' (Grosz, 2012: 54).

2. Note the similarity here with the emergence of the psychic: there is the emergence of a problem due to a disparity with an externality that then requires resolution in a higher concurrent systemization.

3. A recent contribution to the field of media studies that undertakes a psychosocial approach is *Psychosocial Explorations of Film and Television Viewing* (Whitehouse-Hart, 2014). Although very promising, it ultimately fails to deliver a thoroughly psychosocial approach, instead attempting a syncretic union of theories from both psychology and sociology. This approach runs into difficulty when describing phenomena such as emotion, which remains purely psychological rather than co-constituted with the social. What's more, it tends to rely on traditional theories of ideology for the hylemorphic application of identity formations, thus bringing together incongruous theoretical approaches.

4. The phrase 'ontogeny of behavior' describes the idea that 'behaviour could be treated like a body' (Merleau-Ponty, 2003: 148). That is to say that the behaviour of a body can be ontogenetically accounted for in the same way as its form.

5. We can correlate this stage of the image-cycle to von Uexküll's theory. The achievement of a 'real mental world' is similar to von Uexküll's *Gegenwelt*, the more advanced form of *Umwelt*, which acts as a symbolic mirror of the world to which the organism stands opposed.

Chapter Five

Philosophy of Technology

Before looking at Simondon's philosophy of technology I first want to outline some of the key work in this field with which I will contrast Simondon's position. A good place to begin is to ask why such a philosophy is required. What is it about technology that it requires independent consideration alongside other traditional philosophical areas such as ethics, metaphysics, aesthetics and politics? There are some who deny it has any such claim and subsume technology's relevance to other more fundamental areas, the most simplistic of which would be those who understand technology as being nothing more than a means to an end. An example of this perspective is Frederick Ferré's definition of technology as the 'practical implementations of intelligence' (Ferré, 1995: 26).

Such instrumental perspectives understand technology in terms of its relevance for human situations. Although this definition doesn't rule out non-instinctive tool use in other species, due to its relative unsophistication (for example, lack of application of scientific thinking) it tends to be absorbed into humanistic accounts or ignored altogether. Thus from these perspectives technology is subsumed into other more fundamental humanistic modes of understanding such as its value for the common good (utilitarianism), effectiveness at solving problems (instrumentalism) or as socially constructed as in nominalism, in which denominated technological objects are understood only by those properties that are seen to be appropriate for the conceiver (culturalism). What is at stake for a philosophy of technology, and what these positions don't address, is in what way technology can be understood to *exist* apart from the human.

Philosophy of technology has generally been associated with a group of twentieth-century philosophers who saw in the widespread development and implementation of industrial technology not just the means to human ends but a phenomenon with its own power that in some way stood apart from humanity and even nature, that is to say that it had its own essence. Of this group of first-generation philosophers of technology I will briefly discuss two of the most prominent: Heidegger and Ellul.

FIRST-GENERATION PHILOSOPHY OF TECHNOLOGY: HEIDEGGER

In his influential essay 'The Question Concerning Technology', Heidegger (1977) also begins his analysis of technology by considering it from instrumental and anthropological perspectives, and although he sees these as being 'correct', he also argues that both approaches fail to uncover the 'essence' of technology.

To uncover this essence requires further investigation into the correctness of the instrumental definition to understand 'what is the instrumental itself?' (Heidegger, 1977: 6). This enquiry reveals that what is really under discussion is the nature of causality. That is, instrumental talk is a way of referring to means and ends, which are causal terms. But what kind of causality, Heidegger goes on to ask, is instrumentality? His reply is that instrumentality is a *way of revealing*, a way of bringing-forth, and most significantly, 'It is the realm of revealing, i.e. of truth' (Heidegger, 1977: 12).

The connection Heidegger makes between revealing and truth emerges from his understanding of the Greek word *techne*, which is also connected to *episteme* in that both words refer to knowledge: 'Both words are names for knowing in the widest sense' (Heidegger, 1977: 13). Thus more significantly than technology being a pragmatic means to an end (as in manufacturing) is that it is also a mode of revealing, not just in the sense that making something is to reveal something that was not there before, but in the epistemic sense that it is the condition under which understandings and knowings of Being arise. Additionally, Heidegger argues that modern technology, which is indissociable from the projects and understandings of the physical sciences, results in a different kind of revealing to that of preceding technology.

Where older technology is connected with *bringing-forth* in acts of *poiesis*, that is, through artisanal and artistic creation, modern technoscience involves a *challenging* of nature. This challenging refers to the ways that modern technology goes beyond working in harmony with nature *as it is found* and requires its constant reorganization and stockpiling as resource for utilization in human projects. Modern technology is concerned with dominating nature as efficiently as possible for rational human ends with little value placed on other concerns. In one famous example Heidegger describes a dam on the Rhine as reducing the great river to a mere component of a power plant:

> What the river is now, namely, a water power supplier, derives from out of the essence of the power station. In order that we may even remotely consider the monstrousness that reigns here, let us ponder for a moment the contrast that speaks out of the two titles, 'The Rhine' as dammed up into the *power* works, and 'The Rhine' as uttered out of the *art* work, in Hölderlin's hymn by that name. But, it will be replied, the Rhine is still a river in a landscape, is it not? Perhaps. But how? In no other way than as an object on call for inspection by a tour group ordered there by the vacation industry. (Heidegger, 1977: 16)

In the passage that follows there can be little doubt that Heidegger's conception of what constitutes modern technology is not just thermodynamic but also cybernetic. Technology in the modern age certainly exploits the natural world as a supply of raw material and power for production but just as significant is its organization via control and regulation. That such control and regulation have come to direct all aspects of modern life reflects the now global scope of cybernetics.

Furthermore, Heidegger describes the way in which modern technology, in its relation to the development of modern physics, enables the reduction of nature to a 'reporting' of itself as a system of information. The relationship of cybernetics with modern science (technoscience) is understood by Heidegger as a self-sufficient, positive feedback loop in danger of eluding control. It seems to be the combination of this reduction and the reorganization of nature by technoscientific planning to which Heidegger particularly objects. A windmill may utilize the natural power of wind yet it leaves that power unaffected and works within the limitations of its natural occurrence, whereas a damming project reorganizes the flow and route of the river in such a way the water becomes a quantifiable stockpile for future energy production within a wider technical system.

It is this circumscription of natural phenomena within broader technical systems that Heidegger refers to with the term *Gestell (enframing)*, which is also 'that way of revealing which holds sway in the essence of modern technology and which is itself nothing technological' (Heidegger, 1977: 20).

This is to say that *enframing* describes the technological mode of truth (revealing or 'un-concealment', *a-letheia*) that is the way that actuality is constrained to be a 'standing reserve' for use for rational ends. It is *enframing* that Heidegger identifies as the essence of technology even though it is itself 'nothing technological'. Although this *essence* physically emerged in the eighteenth century with the development and use of industrial and scientific technology, it was also present and theoretically dominant in the seventeenth century because of the widespread acceptance of mechanistic causality, thus,

> in Enframing, that unconcealment comes to pass in conformity with which the work of modern technology reveals the real as standing reserve. (Heidegger, 1977: 21)

For Heidegger truth is the manner in which the world is revealed for us, that is, how it appears to us and how we understand it. To understand it as mechanistic and fully calculable—that is, as informational—does not require any actualization of industrial technology. But with such actualization comes the great danger not only of the destruction of natural phenomena but that man will also come to understand himself as mere standing reserve at the disposal of technological systemization, as one among many beings-in-reserve.

This introduces a historical dimension to Heidegger's account in which the essence of technology is understood as a 'destining of revealing', that over time this particular technological enframing not only becomes the dominant way of understanding the real but the horizon of the possible ways by which humanity engages with Being:

> What is dangerous is not technology. There is no demonry of technology, but rather there is the mystery of its essence. The essence of technology, as a destining of revealing, is the danger. (Heidegger, 1977: 28)

The threat then is that the possibility for Being to disclose itself to man in any other way than as standing reserve, due to this event of technological enframing, becomes more remote over time, leading to a world in which man is increasingly integrated into technological systems as a resource with no other understanding of his being than through instrumentality. For Heidegger this onto-destinal process endangers humanity's essence as it impedes it from entering 'into a more original revealing and hence experience the call to a more primal truth' (Heidegger, 1977: 28).

This is not to understand this process as being one of causal technological determinism, for Heidegger is not claiming that certain technological developments cause particular cultural outcomes. Rather, he claims that humanity's place among beings is determined by a specific metaphysical enframing due to modern technoscience.

Subsequently, Heidegger's solution for escaping enframing is also metaphysical in that he suggests what is needed is the development of a 'free relation' to technology. By this Heidegger means the development of an understanding of what the essence of technology is so that it's no longer seen as fundamentally important but 'remains dependent upon something higher' (Heidegger, 1966: 54, quoted in Feenberg, 1999: 185). The something higher is Being and what the 'free relation' to technology enables is that man enter into a relationship with Being that reveals the essence of another mode of truth. That is, it is a relation in which the revealing-concealing nature of technology is understood, thus enabling the disclosure of truth.

This essence is of a truth in which Being is not experienced as a calculable standing reserve but is known via meditation, art and poetry. That's not to say that technology should be dispensed with. Heidegger is not motivated to argue for a world without technology, only that our relation to it needs to change and through that change we will develop to use more artisanal technology (such as the windmill described earlier) and, presumably, avoid the runaway reinforcement of technoscientific enframing because of being in thrall to its self-determined 'successes'.

One of the reasons Heidegger's philosophy of technology remains so influential is that it draws together some of the more traditional ways of thinking about technology even if ultimately going beyond them. He accepts both the instrumental and anthropological descriptions of technology as 'correct' even though they ultimately fail to describe technology's essence. He also acknowledges that technology is a historical phenomenon, emerging from as far back as ancient Greece. His position also understands technology ontologically in the sense that in its modern form it substantially stands in opposition to humanity and nature in a way that operates differently to Aristotle's traditional modalities of causation, replacing these with a more cybernetic account of goal-directed positive feedback. As a mode of revealing Heidegger is clear that the essence of technology is also epistemologically conditioning, a situation that ultimately has ethical consequences.

What Heidegger's account also has in common with other so-called first-generation philosophers of technology is that he also describes man's relation to technology as involving alienation. His description of alienation, as we will see, is different to those given by both Marx and Simondon. For Heidegger technology is alienating because it distances man from the essence of truth. As we have seen, this is because technology both provides its own configuration of truth (that is, the way that beings manifest themselves, as standing reserve, for example) but also conceals man from his own essence by doing so.

What ultimately enables such a substantive account is that Heidegger's account is transcendental in the sense that it describes the conditions of possibility for the revealing of beings as governed by modern technology and understands this as what is essentially important about it. Accordingly in this account heterogeneous instances of actual technology have little import other than as symptoms of the deeper problem of the technological condition in which mankind finds itself, hence the surprising ability for Heidegger to understand concentration camps and industrial agriculture as part of the same historical process of enframing.

JACQUES ELLUL

Another influential first-generation philosopher of technology is Jacques Ellul. He also has a cautionary outlook regarding modern technology, warning about what he sees as its determining drive for ever-greater efficiency, that is, *operational effectiveness*. Although arguing that he develops his theory by way of empirical investigation, his conclusions are not so far from Heidegger's in that he sees mankind as transcendentally conditioned by *technique* or 'technological civilization' in much the same way that Heidegger argues that mankind is captured by enframing.

For Ellul *technique* is defined as 'the totality of methods rationally arrived at and having absolute efficiency (for a given stage of development) in every field of human activity' (Ellul, 1964: xxv). As such *technique* is not understood principally as a form of unveiling. Perhaps due to his more sociological perspective, Ellul doesn't consider *technique* as an essence but as a conditioning influence that invades every aspect of the life of Western civilization.

Ellul understands *technique* as forming a system that has a structuring effect on human society to such an extent that its goal can be seen as the removal of human interference within that system. Explicitly referencing cybernetics when describing automatic piloted aircraft, he writes:

> This progressive elimination of man from the circuit must inexorably continue. Is the elimination of man so unavoidably necessary? Certainly! Freeing man from toil is in itself an ideal. Beyond this, every intervention of man, however educated or used to machinery he may be, is a source of error and unpredictability. (Ellul, 1964: 136)[1]

Ellul identifies the root of the problem in a reductionism towards calculation and automation, the perfection of which sees mankind as an impediment. Like Heidegger he also describes the roots of *technique* as emerging in ancient Greece and flowering during the Industrial Revolution when it exploded 'in every country and in every human endeavour' (Ellul, 1964: 42). Ellul is even more pessimistic than Heidegger, however. Due to its metaphysical nature the possibility for a different relation to technology is contained in Heidegger's theorization of it. However, Ellul doesn't understand technology in terms of a relation to truth. For him technique is truly determining of culture in that since the Industrial Revolution it has become self-augmenting and develops 'without decisive intervention from man' (Ellul, 1964: 87). This has occurred due to the widespread penetration of technique into all aspects of cultural life, meaning that technology now structures it to such an extent that any change in technology necessitates a cultural transformation. As such Ellul's cyberpositive theorization doesn't contain a redemptive aspect and he remains pessimistic about the chances of escape from this situation without some form of divine intercession.

CRITICAL THEORY OF TECHNOLOGY

Another significant current running through philosophy of technology is the critical theory of the Frankfurt school. I will here briefly mention the position of Herbert Marcuse who understood technology as symptomatic of a greater structuring force, that of instrumental reason.

Although Marcuse understands technology as having a formative role socially, particularly that of subordinating humans and nature to the ends required by systemization, he doesn't essentialize this as Heidegger does, and thus, although emphasizing the same functionalist tendency towards determination as Ellul, he offers a challenge to the latter's understanding of this as having passed the point of no return. Instead Marcuse understands technology as one way that society is organized through power and domination, thus opening up the possibility of the creation of other kinds of technologies that support different social values that aren't so instrumental and exploitative. It is through this opening that Andrew Feenberg utilizes Marcuse's thought to help establish his social constructivist project. This constructivism necessarily counters determinism by denying the cybernetically positive dynamic of technology. In doing so it also avoids essentializing technology.

For now I just want to highlight that aspect of what might be called the first generation of thinkers of modern technology that involves, to some extent, thinking technology as having a structuring effect on reality, both epistemologically (in that it reveals) and ontologically (in that it determines). With all of the thinkers considered thus far the essence of technology or *technique* is equated to a self-amplifying operation of control and efficiency via rationalization, measurement and calculation. These thinkers see technology as both symptom and perpetrator of a more fundamental operation that stands ontologically opposed to an essential humanism. In each case it is a return to a form of humanism that is seen as the cure: Heidegger argues for a more poetic relation between man and world, Ellul vainly seeks the answer in both Christian spirituality and left-wing politics, both of which he sees as having themselves been compromised by the very *technique* he hopes they might cure. Marcuse and Feenberg both propose a form of constructivism whereby ethical values are materialized in technology.[2] As we will see, Simondon is also concerned with these issues, although he develops a different response.

SECOND-GENERATION PHILOSOPHY OF TECHNOLOGY: EMPIRICAL PHILOSOPHY OF TECHNOLOGY

It's necessary to mention a subsequent group of philosophers of technology to those discussed above who can be understood as constituting an empirical turn in that they 'look at concrete empirical manifestations of different technologies' (Achterhuis, 2001: 3). This turn denoted a shift from the essentialist and transcendental approaches that were the hallmark of the first-generation thinkers and instead investigated concrete instances of technology in terms not only of the various transformative effects they have on the world but also in order to understand the heterogeneous aspects of social reality that shaped their development.

Thus unlike those earlier philosophers who give a broadly linear causal description leading from technical essence to technical society, these philosophers attempt accounts that allow for a more co-constitutive relation between technology and society. As such many of the thinkers who fall under this description have strong links with social constructivism.

I won't describe these thinkers here in depth as we are here focused on Simondon's work but some key names that can be associated with the empirical turn are Bruno Latour, Andrew Feenberg, Donna Haraway and Don Ihde. Each of these thinkers has quite different ways of articulating the relationships between technology, society, culture and nature but each of them sees the relationships as co-constructive in some manner. What I will argue in this and the following chapters is not only that Simondon develops a unique philosophy of technology that directly responds to those broad questions regarding the challenge of modern technology we witnessed with Heidegger and Ellul, but additionally that Simondon also develops his thinking in an empirical manner that considers the concrete being of technical phenomena in a way that significantly predates those philosophers associated with the empirical turn.

However, and crucially, Simondon charts a philosophy of technology that navigates a course that can't be subsumed to any of these positions. He avoids essentializing technology, although does argue that it has its own mode of existence. This modality means that he can avoid constructivism without lapsing into determinism. Simondon is able to achieve this due to the unique place that technology occupies across and between the physical, vital and psychosocial individuation we've outlined in previous chapters.

SIMONDON'S PHILOSOPHY OF TECHNOLOGY

Simondon's philosophy of technology is principally developed in his 1958 book *Du mode d'existence des objets techniques (The Mode of Existence of*

Technical Objects) as well as in some other shorter texts that have been collected in the volume *Sur la technique* (2014). This work complements and builds upon the work on individuation and is a further application of his relational ontology. In fact, technology holds a special place in his thought, as it is the domain that traverses all three of the regimes of individuation that I have described as constituting nature for Simondon. It also plays an important role in his account of social transformation.

Unlike Heidegger and Ellul, Simondon is able to avoid essentializing technoscience as a self-generating and cybernetically positive dynamic due to the ontogenetic nature of his realism. That is to say that for Simondon the generative development of technology is not determined by an essence because although it has a mode of individuation this modality involves the admixture of all three regimes of individuation. To make this explicit it will be necessary to first understand the modality of technicity that Simondon describes and then clarify its relation to both nature and culture. By doing so we will more fully be able to comprehend the epistemological importance of technology as well as the importance it has both aesthetically and axiologically.

CONCRETIZATION AND THE ASSOCIATED MILIEU

According to Simondon the development of technical lineages should not be understood through functional or instrumental progression (for example, interpreting the history of recording devices as a lineage) but through the development of internal operations. Thus steam trains are not of the same lineage as electric trains, even though they fulfil the same function, because their actual technical mechanisms have developed from different origins. This already disables a species of essentialism that settles essences around a homogeneous functional lineage.

Instead technical development should be understood as one that is led by the technical structure itself, which in the course of its operation unveils and concretizes previously undiscovered synergies and relationships. This mode of existence is the process of a technical object's development via the notion of concretization, which can be understood as a directed and unifying transduction[3] within the regime of physical matter. Simondon describes a concrete technical object as

> one which is no longer divided against itself, one in which no secondary effect either compromises the functioning of the whole or is omitted from that functioning. . . . The essence of the concretization of a technical object is the organizing of functional sub-systems into the total functioning. . . . Each structure fulfils a number of functions; but in the abstract technical object each structure fulfils only one essential and positive function that is integrated into

> the functioning of the whole, whereas in the concrete technical object all functions fulfilled by a particular structure are positive, essential, and integrated into the functioning whole. (Simondon, 1980: 31)

The shift from abstract to concrete is key here. The abstract form of a technical object is one that has 'an intrinsic perfection of its own that needs to be constituted as a closed system in order to function' (Simondon, 1980: 14). Such abstraction describes the object as being hypertelic in that being closed it is abstracted both from its genesis as well as the possibility for further pluri-functional development.

Simondon uses the example of the move from water-cooling to air-cooling systems in combustion engines to demonstrate the shift from abstract to concrete. A water-cooled engine contains two abstract systems that are each perfectly suited to carrying out their own specific functions, the engine itself and the cooling system. However, when these two systems are linked together there is a degree of disparity between the operations of each. Simondon refers to the joining of conflicting technical systems as creating a 'series of problems to be resolved' (Simondon, 1980: 14).

The concretization process resolves such problems so that the system as a whole no longer operates with a level of disparity. Thus the development of air-cooled engines, by the addition of gills to the cylinder block, is seen as a measure of concretization because the engine's cooling function is no longer provided by a separate closed water-cooling system requiring its own operating conditions that conflict with the operation of the engine. Instead the cooling occurs as part of the normal operation of the integrated technical system. Additionally, a further degree of concretization can be discerned because the same gills that are used for air cooling also act as structural supports for the cylinder head. We therefore witness in this progression a move from conjoined abstract structures, which are problematic, to a single pluri-functional concrete system.

Another aspect of the definition of a technical object is that, as part of the organization of functional subsystems into a total functioning, an associated milieu is also invented and maintained. It is important that the specific meaning Simondon gives to the term *invention* is grasped here. Invention does not refer to the traditional hylemorphic notion in which a subject has an idea and then builds something that corresponds to that idea; rather it is the 'birth' of a new environment or 'regime of functioning' (Massumi et al., 2010: 39) brought about by the operation of recurrent causality involving the actual operation of the technical individual itself.

Simondon is not claiming that technical invention occurs without the aid of an inventor, rather that the recurrent physical causality also plays an important role for it is this which is both guided into existence by the inventor, as well as itself guiding the inventor. An invention can be strongly emergent in that the nature of the resulting regime of operation is not wholly foreseeable. In such cases the inventor's role is like that of a midwife helping the new regime of operation come into being. The invention occurs because a jump is made, which is both ontological and usually also epistemological, and this jump is justified by the relationship that is instituted within the environment it creates (Simondon, 1980: 59). One of the qualities of allagmatic epistemology is that to understand an individuation is to individuate one's thought along with the individuation of being, to think being in its ontogenetic becoming:

> Beings can be known by knowledge of the subject but the individuation of beings can be seized only by individuation of the knowledge of the subject. (Simondon, 2013: 36)

With invention the inventor's role is to simultaneously think the solution of problems along with attempts to actualize it. One way to think about this is the process of iteration that inventors undertake in that initial work on inventions usually don't operate as expected but reveal new material potentials that can be folded back into the inventive process. In such instances the technical objects are co-constitutive in the inventive process with the ideas of the inventor. The description of invention related to technical objects has several similarities to that describing both physical and vital individuation. In reference to the latter, Simondon writes:

> The state of a living being is as a problem to resolve which the individual becomes the solution to through successive arrangements of structures and functions. . . . The development may then appear to be the successive invention of functions and structures that solve, step by step, the problem carried internally as a message by the individual. (Simondon, 1964: 223)

Just as the progression of individuation in the living involves the resolution of disparities by an invention that is constructive and 'incorporates the poles of the disparity that is the problem' (Simondon, 1964: 241), what we find in the development of the technical lineage is the progressive resolution of problems through concretization. As such the concretization of the technical object is also, in a sense, its naturalization.

Additionally, just as with physical and vital individuation, Simondon is clear that technical invention requires that any problem that is resolved results in the individual being partly constituted in a relationship with a new environment and not just with the abstract individual being 'added to'. This new environment is what is named the technical individual's *associated milieu*.

> Such individualization is possible because of the recurrence of causality in the environment which the technical being creates around itself, an environment which it influences and by which it is influenced. This environment, which is at the same time natural and technical, can be called the associated milieu. By means of this the technical being is conditioned in its operation. This is no fabricated milieu, or at least it is not wholly fabricated; it is a definite system of natural elements surrounding the technical object. The associated milieu is the mediator of the relationship between manufactured technical elements and natural elements within which the technical being functions. (Simondon, 1980: 60)

Of particular importance in reading this passage is how the word *natural* is understood. As we have seen in Simondon's ontology, nature consists of the pre-individual, which individuates in three regimes: the physical, the vital and the psychosocial. Therefore it is possible to interpret this passage as stipulating that a technical object's associated milieu can be constituted in relation to any of these regimes. However, in *The Mode of Existence of Technical Objects* the description of concretization presented gives an overwhelming impression that by 'nature' in this passage Simondon is limiting his scope to the regime of physical individuation. This impression is emphasized by the disdain with which Simondon greets the intrusion of cultural factors into technical concretizations—this can be witnessed in his scorn for decorative fins, power steering and starter motors for automobiles, which he explains away as advertising-driven gimmicks—which either add abstraction and disparity to a technical individual or disrupt its concreteness. Such abstraction entails that the automobile runs the risk of becoming hypertelic: 'The automobile, this technical object that is so charged with psychic and social implications, is not suitable for technical progress' (Simondon, 1980: 21).

The concretization process, which Simondon describes as the true evolutionary principle of technical objects (which he also calls mechanology), thus also operates separately to economic and cultural concerns and can't be reduced to 'anterior scientific principles' (Simondon, 1980: 48). It is this study of technical individuals that aids the discovery of synergies, boundaries and indetermination in their operations, which lead to the possibilities for further invention.

Mechanology is also framed as a type of scientific development in that it reveals previously hidden virtualities and makes them available for further concretization. There is thus, to some extent, a resemblance between mechanology and technoscience. As Bernard Stiegler writes:

> If a mechanology is necessary, this is because the laws of physics, no more than those of sociology or psychology, or all of these as a whole cannot suffice

> to explain the phenomenon of the technical object qua the genesis of an individual and production of an order. (Stiegler, 1998: 76)

This account of mechanology as a scientific investigation emphasizes the distance at which Simondon keeps the process from cultural influence. Because of this distancing it is tempting to see mechanology as a process of purification such as described by Latour (1993) in *We Have Never Been Modern*; that is, as a process that constructs an account of nature purified of any social involvement and vice versa. This sense of mechanology as a purified process can also be discerned in Paul Dumouchel's summation of Simondon's position:

> It is not because of the uses we put it to that modern technology radically transforms the world, but because technology gives existence to phenomena which were not there before and because technical individuals provide the conditions of the processes which constitute them. Thus there is no alternative technology which contains different values with respect to nature. What technology teaches us is that there is no 'nature' in the sense of a set of events and processes which are essentially different from those which are produced artificially. According to Simondon there is no technology which can respect what is, for technology is essentially the coming into existence of the virtual. (Dumouchel, 1995: 268)

Dumouchel makes clear that any leap of invention requires the existence of the technical object. There is a sense in which the technical object comes first and transitions occur around it. But these developments progress through an internal logic divorced from the normative domain, that is to say the cultural. But how is this possible if, as we have seen, in the account given in *Imagination et invention* the invention of the technical object involves ideation?

One way we can understand Simondon's continuing assertion for the purity of the technical mode of individuation is to make clear that the causality of this operation is understood as not necessarily determined by cultural influence; that is, that it is a particular mode of physical individuation with a certain amount of *autonomy*.

Given Dumouchel's assertion above that technology doesn't mark out a set of 'events and processes' that differentiate it from nature or indeed that there is no alternative technology within which inhere alternative values, we can already discern a sharp distinction from the first generation of philosophers of technology such as Heidegger, Ellul and Marcuse. However, Simondon's position isn't as straightforward as Dumouchel describes. To understand why I will first turn to the claim Simondon makes for a distinction between the natural and artificial before more fully explicating his understanding of the relation between technicity and culture.

NATURAL AND ARTIFICIAL

Although Dumouchel is correct regarding the indiscernibility of the natural from the technical at a fundamental ontological level, for Simondon the technical object can help distinguish a clear difference between the natural and the artificial. Simondon explains this distinction using two examples. One is the concretization of a technical object that, although initially developed to operate within the conditions of a laboratory, progresses to be able to operate outside of this regulated environment. The second example is that of a plant that is only able to survive by being cared for in a greenhouse.

In the first example Simondon argues that the concretizing process, which enables the technical object to operate independently of the laboratory conditions on which it was originally reliant, is akin to a process of naturalization in that the object develops from a primitive and artificial state to 'more and more resemble a natural object' (Simondon, 1980: 47). By this Simondon means that the technical object has become more autonomous in its operation and can regulate itself dynamically in regard to its relations with other objects and environment. Because of the autonomy that concretization gives technical objects Simondon argues that we can go so far as to understand them as natural through analogy:

> Because the mode of existence of the concrete technical object is analogous to that of a spontaneously produced natural object, we can legitimately consider them as natural objects; this means that we can submit them to inductive study. (Simondon, 1980: 48)

The naturalness of an object is therefore not related to its origin or regime of individuation but to the extent to which it operates in conjunction with its environment. As we have seen, a technical individual with an associated milieu creates, in part at least, the conditions necessary for its own operation. To the extent to which it does so, it can be said to be naturalized. The more open this technical object is to a broad range of environmental conditions the more naturalized it can be understood to be. To say it is naturalized is not, of course, to say a technical object is therefore vital.

Such naturalized objects are worthy of 'inductive study' due to the possible unique potentialities their operations may reveal. Simondon also warns against reductionism by asserting that it is through empirical study (mechanology) that the mode of operation of technical objects can come to be understood, that is, by individuating one's knowledge with that operation and not applying prior knowledge to it in order to explain it.

The second example discussed is of a plant that has been nurtured in a greenhouse environment on which, along with the intervention of people, it relies for its ongoing survival and reproduction. For Simondon this is an example of *artificialization*. Despite the organic origin of the plant Simondon points to the fact that its natural status of having an integrated pluri-functionality in relation to its environment has been destroyed, thus rendering it artificial:

> The essential artificiality of the object resides in the fact that man has to intervene in order to keep the object in existence by protecting it from the natural world and by giving it a status as well as existence. (Simondon, 1980: 46)

As the plant becomes more reliant on the 'artificial regulation of the greenhouse' it is likely it will become more and more *abstract* as it can be manipulated so that rather than being a unified system of coherent functions, individual functions (for example, flowering, bearing fruit) can be enhanced through human intervention to the neglect of others. In such cases we can understand the abstract not as the opposite of the concrete but as a specific domain of the concrete that has been emphasized, thus causing a dephasing.

In 'Culture and Technics' (2015) Simondon describes the agricultural heritage of the term *culture*, including how husbandry, in working on emphasizing certain qualities of animals, leads to 'artificiality through adaptation'. That is to say that by intervening in this way the species' initial adaptation to its environment is broken, if not at least warped. The breaking of this relationship with its milieu not only leads to the adapted species becoming reliant on mankind to survive, but also to 'hypertelic dysfunction' because in being bred for narrow functional concerns (for example, milk production, meat) the species is likely to lose other functions such as the ability to independently reproduce.

It is not the case, however, that through increasing concretization the technical object might ever become vital, nor that a plant rendered artificial through growing in a greenhouse could exist as non-vital. For an individual to be vital requires both the maintenance of itself as an ongoing 'theatre of individuation' (Simondon, 2013: 27) as well as a certain topological relation between interiority and exteriority. Simondon maintained that technical objects are not capable of these operations, which is one of the reasons why the cybernetic analogy of humans with automata is mistaken.

Similarly, a living plant, however artificial, is still individuating in the vital mode. What Simondon deems as natural then is closely related to that entity's level of operational autonomy in relation to the environment within which it is situated. It requires a level of concretization and openness to an environment, which enables an ongoing autonomy, and if necessary, invention as opposed to hypertelic closure.

It is worth revisiting Dumouchel's quote again to question whether Simondon's view of technology actually leads to his conclusion that 'there is no alternative technology which contains different values with respect to nature' (Dumouchel, 1995: 268). Much depends on how the natural is defined and Simondon's distinction of natural from artificial indicates that Dumouchel must be mistaken when he claims:

> According to Simondon there is no technology which can respect what is, for technology is essentially the coming into existence of the virtual. (Dumouchel, 1995: 268)

Simondon's discussion of the natural and artificial indicates that he does attach a greater value to autonomous and concrete individuals described as natural over abstract and artificial ones. As we will see this distinction is at the heart of his aesthetics and also enables him to make judgements about different technological instances. I will bracket this concern for now and continue my discussion by contrasting Simondon's notion of concretization with how it has been utilized by the critical theorist Andrew Feenberg.

ANDREW FEENBERG'S HUMANIST ACCOUNT OF CONCRETIZATION

Though by no means asserting concretization as a democratic socialist theory, Andrew Feenberg uses it to support the political idea that 'socialist demands for environmentally sound technology and humane, democratic, and safe work are not extrinsic to the logic of technology but respond to the inner tendency of technical development to construct synergistic totalities of natural, human and technical elements' (Feenberg, 2002: 188).

His proposal is for a concretization whose scope is expanded beyond Simondon's to include within its operation the aims of critical politics. It is then a technosocial form of concretization. This is one in which social norms and constraints are embedded into technology, from which they are often forgotten or even assumed to be part of the object's 'inevitable technical destiny'. For Feenberg social values are another area of virtuality that can be concretized into technical objects in what he describes as a 'technological unconscious'.

The inclusion of these concerns into the operation of technology does not require that these technologies need become less productive. Indeed the social codes incorporated into any technology could just as well be capitalist in nature as critical. Importantly, Feenberg maintains that the choice of social codes concretized into technology is essentially a political choice and is evidence of technology's underlying ambivalence.

For Feenberg technological systems help structure our everyday life but are open to concretization according to a different trajectory than that supplied by contemporary capitalist operations, through the condensation of more social aspects such as an appreciation of 'workers' skills, human communication, and environmental limits' (Feenberg, 1996) into their actual operational structure.

If there were invariably social aspects involved in concretization this would seem to contradict the purified account of technical concretization that Simondon described above. It is true that if we wished to maintain this purified position, we could just maintain that any such introduction of social concerns into technological development is *de facto* strictly not mechanological. However, this does not mean that we must therefore agree with Feenberg's position that an ethics of technological development be founded on a humanist 'politics of technological transformation'.

As we will see, Simondon is resolutely clear that such an imposition of cultural values onto technology is a problem in need of resolution and not the solution itself. Instead what Simondon will propose is that the operation of technical individuation is not to be subordinated to humanist values but that values will be derived from that operation itself. Before I can address this argument, however, it will be necessary to more fully describe how Simondon understands technology's role in relation to a new form of humanism and encyclopaedism[4] as well as in relation to both Marxism and Capitalism.

HUMAN PROGRESS

So far I have described that part of Simondon's philosophy of technology that investigates the nature of the operation and individuation of technical individuals and how this relates to technical lineages. As I will now explain this is just one aspect of Simondon's account of technicity that, to be properly understood, must be situated within his wider philosophical project.

In his text 'The Limits of Human Progress: A Critical Study', Simondon (2010a) sketches an account of human cultural progress as developing in the same genetic manner as technological development: as the progressive operation of concretizing relations between differing domains as a response to resolving disparities. To be a universal account Simondon states that it must take into consideration 'the entire system of activity and existence constituted by what man produces and what man is' (Simondon, 2010a: 230). As such Simondon's project should be understood as encyclopaedic. To achieve this broad scope he divides these activities and their products into ontological domains (for example, language, ethics, religion as well as the technical). These domains come into problematic relation with one another and produce emergent concretizations. For Simondon, during different historical periods the concentration of human activity occurs in different domains. For example, in ancient classical civilizations there was a concentration of activity on language development whereas in the mediaeval period development mainly concentrated in the religious domain. During the current period, Simondon contends, the concentration of activity is focused on developments within the technological domain.

As with mechanology, development within each domain is described transductively in terms of concretization and saturation. Importantly, progress within any domain occurs between humanity and the concretizations of that domain that have already occurred and with which humanity is said to form a system. The further development of this system requires that it isn't saturated but remains in a state of internal resonance, thus engendering further progression. The saturation of a domain leads to its stagnation. This is because saturation consists in the complete determination of all available 'virtualities' or potentialities for development in any given system. Thus regarding the domain of language at the close of the ancient world, Simondon asserts:

> It became purely a matter for grammarians and formalist logicians seeking etymological rectitude in naming. Surely, a grammar or a formal logic do not reflect man, or at the least reflect only the smallest part of man, one that should not be inflated. (Simondon, 2010a: 231)

THE AESTHETIC DIMENSION OF JUDGEMENT

Though we live in a world currently heavily invested in the development of the technical domain there is no reason to believe that this process will not also reach saturation and the focus for investigation shift to another domain. However, Simondon also notes that the chances of humanity becoming alienated from technological concretizations are less likely than from those of the domains of language and religion because

> technology is even more primitive than religion: it connects with the elaboration and satisfaction of biological desires themselves. . . . Thus there is at least the chance that the seeds of the decentring of man, and thence of the alienation of the objective concretizations which he produces, may be feebler in technology than in language and religion. (Simondon, 2010a: 232)

The importance of technology for addressing biological desires is one of the reasons Simondon argues for the need for a technological culture. Although the development of technology occurs via the progressive uncovering and utilization of potentialities this operation must also be contextualized within this broader account of human progress, which requires an ongoing internal resonance between all domains, as well as 'durable overlappings' between them.

Simondon maintains that this is achieved via reflexive, philosophical thought that, as the 'conscious form of the internal resonance formed by man and the objective concretization', can prevent alienation between man and technology by ensuring that technological progress becomes an 'integral part of human progress, by forming a system with man' (Simondon, 2010a: 235). In this there are echoes of the description of transindividuality as that which involves the co-current individuation of the emotional interior with exterior action.

In *The Mode of Existence of Technical Objects* Simondon's concern is not of the same kind as in his main thesis (*L'individuation*). Where in the latter his goal is to give an ontogenetic account of individuation from the pre-individual to the transindividual, in the former his aim is to describe the genesis of technicity itself as a phase alongside culture. Although in *The Limits of Human Progress* Simondon describes how he understands cultural development as occurring via the shifting of phases in different domains he doesn't describe their genesis in that text, and consequently how they are organized developmentally and how this has normative consequences. Once again, just as with physical, vital and psychosocial individuation, in this text Simondon utilizes the phase-shift as the schema by which 'the temporal development of a living reality proceeds' or, in other words, how the development of the human mode of being in relation to the world progresses.

Unlike with dialectics, Simondon's account doesn't require negation to propel becoming, the engine being the division, and subsequent relation, of phases. Thus rather than the sense of inevitable onwards progression that dialectics describes the schema based on phase-shifts is an account that can describe development as constituted by a number of different phases that fall in and out of balance with one another around a central point. As described in *The Limits of Human Progress* different phases are dominant at various times throughout history but can fall from prominence, dephase or even divide further, depending on the extent of their saturation and relation with other domains. This is therefore a very different understanding of progress to that traditionally associated with modernism, for example.

Crucial for understanding the role technicity plays for Simondon as a function of human thinking is the account of its emergence as a phase of man's relation to the world. Just as in *The Limits of Human Progress*, in which Simondon applies his informational schema to the development of domains of knowledge, in a remarkable chapter in *The Mode of Existence of Technical Objects* called 'The Phase-Shift of Primitive Magical Unity' he applies his ontogenetic, informational theory of individuation to describing how the different functional modes of human thinking that structure the human relation to the world emerged and developed. His overall goal in describing this dynamic is to demonstrate the importance of philosophy for integrating 'technical reality with culture' (Simondon, 2001: 162).

According to Simondon prior to there being any development in the human relation to the world, that is, prior to there being any phase activity pertinent to its relation to the world, humanity existed in a 'magical' mode of being. This mode is *the* primal *vital* mode of man-world relation and describes a mode of being prior to the division of world into subject and object or singular and universal. This magical mode denotes the 'unity of the living and its milieu' prior to any shift that would lead to a distance being cognized between them. The mode of this unity is reminiscent of von Uexküll's description of the paramecium as resting 'in its Umwelt more surely than the infant in its cradle' (Merleau-Ponty, 2003: 170), which is to say that it is undisturbed by problems coming from the environment that may require an inventive response. This mode also aligns with the description Simondon gives of the vital mode of 'simple life' in which the living individual maintains an affective balance between perception and action prior to the entry into psychic existence proper.

Although there is no subject-object division for this mode, Simondon argues that it does have a reticular structure involving privileged places and times that correspond to 'the most primitive and most fecund of organizations' (Simondon, 2001: 164). What Simondon has in mind here is the most basic systemization[5] of the imagination resulting from conspicuous topographic elements such as mountains, hills and rivers and temporal phenomena such as solstices and full moons. Such phenomena are 'privileged' and 'localize and focalize the attitude of the living vis-à-vis its milieu' (Simondon, 2001: 164).

It is this 'initial active centre' (Simondon, 2001: 159) of the magical unity of the man-world relation that is central to the narrative of Simondon's account and from which technicity gains its significance. The magical mode is understood as a unified state from which the phases of culture individuate. It is the dephasing of this primal mode of human-world relation that results in the initial creation of the phases of technicity and religion. These phases describe two different modes of relating to the world, which result from the dephasing of the reticular structuration so that figure and ground become 'detached from the universe to which they adhered' (Simondon, 2001: 167). These phases of technics and religion (and their modes) are the mediating relations between the human and world, a mediation that 'takes a certain density' (Simondon, 2001: 168), that is, a thickening and complexification of spatio-temporal experience.

Simondon understands these two phases at the poles of various relations. Thus the religious phase is concerned with the subjective and focuses on the universal ground of experience as a unity (ubiquity and eternity). In opposition to this technicity is concerned with the objective aspect of experience, its focus is with the figural and singular as previously represented as key points. A technical object is located at a historical time and space, it lacks the quality of universality, it operates by direct contact and deals with the figural rather than the ground. In fact it is one of the qualities of the technical device that as a singularity it can be detached from the ground and operate in multiple places (through duplication) and times. The religious relation, however, is concerned precisely with that which is universal and understood as ground. The technical object lends itself to empirical investigation, finding any ground it encounters as 'anonymous and foreign' (Simondon, 2001: 171). Religious thinking lends itself to the transcendental such that any individual it considers is 'always seized as less than unity, to be dominated by an approached totality that infinitely exceeds it' (Simondon, 2001: 172). The figure and ground in both modes of thought are therefore detached from the world, free to be emphasized without being limited by the other. In the religious mode transcendence becomes detached from the world, standing over it, preceding it, yet unknowable for mere subjects. In empirical technical investigation the figural is detached from the ground, from the milieu from which it develops, and is thus limited in scope for understanding the world.

Just as we saw in the last chapter when describing the difference between individuation and individualization, Simondon once more demonstrates the insufficiency of the transcendental and empirical perspectives for describing the unity required to make sense of a true immanence. This time he does so from the vantage of thinking through the ontogenetic development of the human relation with its milieu that began as a 'primitive complex of man and the world' (Simondon, 2001: 173) prior to dephasing.

The religious and technical phases are linked by aesthetic thought, which acts as a neutral point between them. Importantly for Simondon aesthetic thought 'is not a phase but, rather, a permanent reminder of the rupture of the unity of the magical mode of being and a search for a future unity' (Simondon, 2001: 160).

Following the primary phase-shift resulting in the religious and technical phases Simondon describes a subsequent division of each of these phases into practical and theoretical modes. It is in the gap between the theoretical modes of religious and technical thought where scientific knowledge develops as a relation. Similarly ethics develops as the mediating relation between the practical modes.

Simondon also describes a further phase-shift, which occurs with the development of industrial technology. Where the first shift was productive of empirical scientific naturalism as well as transcendental religious thought, this second shift develops a technical thought that has turned to elaborating the human world, plus a corresponding development in the religious domain that produces 'the major global political movements which are properly the functional analogue of religions' (Simondon, 2001: 214) and that attempts to think humanity in its totality.

In his account Simondon explicitly ties the progression of the phases of human relation to the world with that of the development of technicity. In the first phase technology, in its artisanal form, is opposed to religious thought.

With the second phase shift it is the development of industrial technology, which has a more reticular structure, which provokes the development of social and political thought in the corresponding religious phase. The reticular structure of industrial technology leads to a reformation of the sense of unity from that of the absolute, transcendental unity of the religious to an ecumenism provided by the reticular structure of industrial technology, which is at once local and universal. As such Simondon is hopeful that this new reticular unity, which does not reside outside of individuation (it is more properly immanent and not transcendent), more closely resembles the original magical mode of relation.

This description of phases also has a chronological dimension such that at different moments various phases may be in or out of phase with one another or shifting to new phases. It is the driving motivation for Simondon's work on technology to give an account of how industrial technology is out of phase with culture and how it is the role of philosophy to help rebalance them. It is aesthetic thought that plays a key part in this regulative role as it is more primitive than both scientific and ethical thought, which are situated further from magical unity because they are products of the additional division of religious and technical phases into practical and theoretical modes.

As described in *The Limits of Human Progress*, Simondon understands technicity to be more primitive than religious thought as it is more closely situated to biological desires. Similarly, the aesthetics of technicity will play a significant role for cultural regulation as it is situated closely to the initial mode of magical unity.

To more fully understand how Simondon's genetic philosophy produces a novel way of thinking technology, both universally as well as in its singular instantiations, it is worth briefly pausing my account here to recap in order to start to tease out the ways by which he navigates a path between the transcendent and the empirical, and invention and regulation. Although Simondon retains a general optimism regarding technology, unlike Heidegger and Ellul, this is not to say that he is uncritical of its theoretical and practical implementations. For example, he's aware of the limitations of the reductive and fragmentary nature of technical schemas of thought, such as Cartesian mechanism, which prompts him to assert:

> The application of schemas drawn from technics does not account for the existence of the totality, taken as a unity, but does account for the point by point and instant by instant functioning of that totality. (Simondon, 2001: 175)

As such he understands technical schemas of thought as running the risk of universalizing that which does not take account of the reality of a unity due to their proceeding inductively from the particular to the universal, from the finite to the infinite. That is to say, no one schema *should* achieve abstract perfection or become 'hypertelic' as it can never account for the fullness of a domain. The same applies in the practical domain in which technical approaches to ethics such as utilitarianism, which divides the unity of life into moments, which can be analysed via a utility calculation, are also guilty of reductive abstraction.

As exemplified in *The Limits of Human Progress* Simondon is also concerned with a more universal account of technicity's position within the human-world system of the totality of functional operations of individuation.

Although, as we saw in the first part of this chapter, Simondon is emphatically concerned with the operational reality of concrete technology, this is but one aspect of a broader analysis concerning humanity's mode of relation to the world. At the very least Simondon's universal diagnosis is for the requirement of a balancing of the various phases he describes, which is one main goal of his encyclopaedic project. Thus unlike Ellul he thinks that our relationship with technology doesn't necessarily have to be one defined by efficiency but can be part of a new form of humanism.

In *The Limits of Human Progress* Simondon makes clear that it is the role of philosophical thought to maintain an internal resonance between man and the various domains, and in *The Mode of Existence of Technical Objects* he also charges reflexive thought with the responsibility for the balancing of the totality of the 'genetic ensemble' in order to maintain a unity:

> This is precisely the goal to be reached: reflexive thinking has a mission to redress and refine the successive waves of genesis by which the primitive unity of the relation of man to the world becomes divided and comes to sustain science and ethics through technics and religion, between which aesthetic thinking develops. (Simondon, 2001: 161)

The role of philosophical thought, guided by aesthetics, is to have a regulative effect on human relations with the world. Simondon understands such a project as requiring a genetic encyclopaedism by which this relation can be understood as one of individuation. What is central for him, as described above, is that for the human the aesthetic dimension is the most fundamental as it is this which most ably facilitates the appreciation of unity. Having situated technicity as one phase among others in a genetic development I will now turn to an analysis of the development of technicity and how progress is understood within that phase.

TECHNICAL PROGRESS AND CULTURE

In the introduction to *On the Mode of Existence of Technical Objects* Simondon specifies that the main problem he wishes to address in that work is that of the growing disparity between a reality in which technical objects play a more and more significant part in human life and the ancient culture by which that reality is governed, which 'has become a system of defence designed to safeguard man from technics' (Simondon, 1958: 9).

In the first part of this chapter I described Simondon's theory of the mode of existence of technology as being one involving the progressive concretization of technical objects. It is precisely this developmental progression that has led not only to the increasing individualizing of technical realities but also to a transformation of mankind's relation to them, a transformation to which culture has failed to react, a situation Simondon succinctly describes as

> the reality governed is made up of man and machines; the code is based on the experience of man working with tools; this very experience is both weakened and remote, because those who use the code have not, like Cincinnatus, just left the handles of the plough. To put it simply, the symbol is weakening and the reality is absent. (Simondon, 1980: 7)

While cultural values have hypostasized technology has undergone periods of revision such that it no longer just involves tools but has progressed from technical elements to technical individuals to self-regulating ensembles of technical individuals, that is, broadly speaking, from workshops to factories to networks.

One way to describe this development is as a progression from simple tools and instruments that extend the functionality and perception of the human body to technical individuals, which rather than extending one aspect of the human (or even animals in the case of tractors or windmills for power supply) replace it altogether for the completion of tool-bearing tasks. An example of this is the Jacquard loom that with its punched-card system superseded the handicraft of loom weaving. Following on from this shift Simondon argues that the twentieth century has seen the further development of technical ensembles, which utilize information feedback to regulate systems and networks of technical individuals.

Simply put Simondon's argument is that contemporary culture is out of phase with technical progression, which has led to it holding 'two contradictory attitudes' to technology, either considering it a threat that stands against humanity (as we saw with Ellul) or as neutral matter that is ascribed meaning and value by humanity (for example, Feenberg). Simondon's aim is to demonstrate that both of these positions are incorrect and that there is a pressing need for culture to understand technology in its technicity.

That in the contemporary world culture and technics are out of phase does not mean that technology should therefore be understood as a threat; rather a rebalancing is required. The historical account Simondon gives of the relation of technical progression with culture is one of recurring concretization and dephasing, the results of which were often positive. For example, in the eighteenth century the development of more refined tools and instruments via improved engineering led to an increase in optimism about overall progress. The nineteenth-century development of technical individuals as power sources was also seen as 'not a frustration' as long as they didn't replace man as the primary tool-bearing technical individual. In these eras technicity and culture can be understood as in phase, with technical progress also involving benefits for those who used machinery.

It is when technical individuals began to displace man and make of him a mere spectator or manager that another notion of progress diverges from that of the craftsman, that is, of the 'cosmic' notion of man's progressive understanding and domination of nature. It is at this point that Simondon's account coincides with Heidegger and Ellul's regarding a domineering technological rationality. Although, as Simondon points out, initially this progression was welcomed as a 'general advancement of humanity' (Simondon, 2001: 117), this was in spite of the fact that it was actually a regression for those craftsmen whose livelihoods were usurped or who experienced deskilling through mechanization.

It is in the twentieth century that these dual notions of progress are understood to be out of phase with culture. At the level of human-technical relation Simondon discerns the development of alienation, while also acknowledging along with the first-generation philosophers of technology that there is a certain technocratic will-to-power at play that is exploitative of the natural environment. I will investigate each of these problems in turn.

TECHNOLOGY AND ALIENATION

It is to a great extent in opposition to Marx's theory of alienation that Simondon sets out his own. Both thinkers situate the development of new forms of alienation with the relocation of labour practices from artisanal workshops to factories in which production is mechanized through the use of self-regulating technical individuals operating in ensembles.

Rather unfairly Simondon describes Marx's alienation as fundamentally legal and economic due to it locating the site of alienation (Marx, 1988: 69–89) in the worker's lack of ownership of the means of production. Against this Simondon proposes a more fundamental kind of alienation:

> Beneath this legal and economic relationship with ownership there exists an even more profound and more essential relationship, that of the continuity between the human individual and the technical individual, or of the discontinuity between these two beings. . . . Alienation does not emerge solely because in the nineteenth century the human individual as a worker is no longer the owner of his means of production, whereas in the eighteenth century the crafts-

> man was the owner of his instruments of production and of his tools. . . . It emerges also outside of any collective relationship with the means of production, at a strictly individual, physiological and psychological level. (Simondon, 2001: 117–18)

To reduce Marx's theory of alienation to a matter of ownership and economics is undoubtedly to commit an injustice to a subject that Marx theorized throughout his lifetime. As such it is worth briefly outlining Marx's understanding of the term in order to more clearly see how Simondon's elision of many aspects of Marx's theory points to a lacuna in his own.

My contention is that Simondon's reduction is mainly due to a focus on the concept of alienation as it is developed in Marx's early writing. Even then it is still a harsh evaluation of Marx's theory to say it only concerns ownership. Additionally, I will briefly touch on Marx's later theory, which at times comes surprisingly close to Simondon's own formulation.

My brief exposition is indebted to Amy Wendling's (2009) excellent study *Karl Marx on Technology and Alienation*, which argues that a significant reason for the difference between Marx's early and late theory of alienation is the result of a shift from a humanistic understanding to a thermodynamic one. There is good reason to suppose that it is Marx's earlier theorization of alienation that Simondon has in mind with his criticisms. In this early formulation, which is found in the *Economic and Philosophic Manuscripts of 1844*, Marx sets forth a broadly 'humanist' theory of alienation, strongly influenced by Hegel and Rousseau, which involves 'four main aspects' (Mészáros, 1975: 14).

First, that Man is alienated from nature in the sense that he is alienated from the product of his labour, which, because of his situation as a waged worker, is not his to enjoy:

> The object that labour produces, its product, stands opposed to it as something alien, as a power independent of the producer. The product of labour is labour embodied and made material in an object, it is the objectification of labour. . . . In the sphere of political economy, this realization of labour appears as a loss of reality for the worker, objectification as loss of and bondage to the object, and appropriation as estrangement, as alienation. (Marx, 1988: Section 5)

This relates to the second aspect of alienation, which holds that Man is alienated from his own productive activity as he can't enjoy its results other than in the abstract sense via the sale of the objects produced. This is to say that the mode of capitalist production entails that labour is no longer enjoyed by the worker as a fulfilling activity but is transformed into an activity necessary to meet the needs of physical survival via a wage. The enforced and stultifying nature of this form of labour is also described as alienating in that it prevents the experience of work as a creative and self-actualizing activity.

The third aspect builds on the previous ones by arguing that alienation so described does not concern just the individual but is social in character. Marx uses the concept of 'species-being' to make this point, arguing that although labour may be alienating for the individual undertaking it, it occurs within a world that has been constructed by a society that is also in an alienated situation. As such mankind's essential nature appears to itself as wholly alienated.

Intimately related to this is the fourth aspect of Marx's early theory of alienation, which holds that Man is also alienated from other men. Marx makes this clear when he writes: 'What applies to man's relation to his work, to the product of his labour and to himself, also holds of man's relation to the other man, and to the other man's labour and object of labour. In fact, the proposition that man's species nature is estranged from him means that one is estranged from the other, as each of them is from man's essential nature' (Marx, 1988: 77).

A psychological aspect of alienation called *false consciousness* is also developed in this respect, which involves the misunderstanding by mankind of its own essential being, instead accepting its situation under capitalism as one that is naturally given. It is from this that Marx develops the powerful concept of ideology.[6]

Simondon's criticisms of Marxism seem focused upon this earlier theory of alienation. First, Simondon is scathing of the focus on labour as the prime relation between man and world, as well as its hylemorphic definition as the imposition of form on matter. It is undoubtedly the case that the notion of labour developed in Marx's early writing is influenced by Aristotle in that not only does it have a hylemorphic character but also that such labour is a means to self-actualization, a notion that suggests a level of humanist essentialism that Simondon is uncomfortable with. Wendling also makes clear that Marx's theory of alienation is developed upon the theory of 'just exchange' as developed in Aristotle's *Ethics* (Wendling, 2009: 25–27). Given this Aristotelian influence it is clear why Simondon disputes the Marxist account. However, it should also be clear that Marx's early formulation of alienation shouldn't be reduced to just economism as Simondon does, for Marx's critique is also political. As Combes remarks,

> While it is true that Marx often relies heavily on the analyses of economists, we must recall that he consistently defines his own project in terms of 'critique of the political economy,' which critique aims to make apparent the mystifying character of the point of view of economists. (Combes, 2013: 73)

That alienation is not just economic, even in the early Marx, is evidenced by the recognition of one of its aspects involving the quality of the worker's activity. That is, that it is important for Marx that such activity involves 'self-actualization, undertaken in freedom from physical need' (Wendling, 2009: 16). That the notion of self-actualization may be too Aristotelian and essentialist for Simondon may be the case but it does point to a criticism that is not purely economic.

In Marx's later work, particularly *Capital*, although economics undoubtedly has a determining role in the production of alienation there is also a focus on the role that technology plays. Undoubtedly in doing this Marx comes closer to Simondon's notion of alienation in that he also recognizes the extent to which machines contribute to deskilling and repetitive and unrewarding labour. Significantly the later Marx also moves away from the humanism and hylemorphism of his earlier work towards a more scientific understanding that utilizes thermodynamics and technology. As such Marx comes much closer to Simondon.

In this reworking Marx shifts from understanding labour as the human imposition of form on natural matter, instead describing labour power as one power among others *within* nature. As an activity, then, labour power involves the transference of energy in accordance with the laws of thermodynamics just like any other activity, such as that carried out by machines. As Wendell explains, this thermodynamic understanding of labour means that

> labor changes from a creative endeavour wrought by human spirit on inanimate nature, as conceived in Aristotle, Hegel, Smith, and Locke, into a mere conversion of energy in which nature goes to work on itself. Labor is no longer a spiritual, form-giving activity that infuses matter; it is merely a part of the transformations of a *natura naturans*. In a related change in thermodynamically influenced physiology, the notion of a vital force or animating spirit is progressively eliminated from explanations of human activity. (Wendling, 2009: 61)

One implication of this flattening of all activity by thermodynamic theory is that the humanist understanding of alienation loses its power. Stripped of its creative essence human labour becomes just another force, no more or less significant than that undertaken by machines or animals. The humanist position from which the abstract nature of labour and its alienated dimension is judged is thus forfeited.

Of course, what Marx is attempting in *Capital*, and also by describing labour in this thermodynamic way, is to tease out the contradictions of capitalism using its own logic. As thermodynamics was the scientific complement to the capitalist worldview it is also utilized to draw out the contradictions of this worldview. So just as Marx depersonalizes human labour as mere labour power he simultaneously retains a more humanistic description of it as living labour and uses this to criticize the impersonal nature of the thermodynamic perspective. This retention of the humanist critique of alienation 'locates Marx with the romantic, humanist resistance to the flattening of human beings, their instruments, their products, and nature itself onto an equitable ontological plane' (Wendling, 2009: 63).

Although it is not necessary to detail Marx's criticisms here it is worthwhile noting that although the later Marx comes closer to Simondon with the utilization of thermodynamics, the commensuration of forces that Marx describes in doing so is not something that Simondon would concede. In fact it predates the same kind of issue Simondon finds wanting in cybernetics with its identification of machines and humans. Indeed Simondon's description of alienation, as we will see, is very much based on the differences between man and technology and how they come into relationship.

Although, as we hope to have demonstrated, Simondon's description of Marx's alienation as fundamentally based on economics and legal ownership is reductive, it is hopefully clearer that Simondon's opposition to Marxism rests on its fundamentally hylemorphic character. The very notion of labour that Marx proposes is too hylemorphic to be included in Simondonian ontology and it is undoubtedly true that Marx understood capital to be too overdetermining. What Simondon is attempting to articulate is a completely different paradigm to that of labour, where not only is the operational nature of the human-technical and human-nature relationships primary but, because of this, it has significant implications for the nature of a collectivity that is not based on the hylemorphic understanding of class or species-being.

However, despite this overcoming of the Marxist understanding by Simondon this encounter between the two thinkers raises questions regarding Simondon's thought, such as: What importance does he attach to the political and economic sphere? I will save discussion of Simondon and politics for the next chapter.

It is at the level of the operational relation of the technical individual with the human individual that Simondon's more fundamental form of alienation occurs. As we have already seen, the technical-human relation is, for Simondon, one of the two modes (the other being the religious-human) that is closest to the initial phase-shift from the magical mode. Given the importance of the magical mode for him it is clear why he values this relation as more important than that of an economically derived alienation.

It is not then a question of ownership that is fundamentally important but that 'man can be coupled with the machine as one equal with another, as a being who participates in its regulation, and not only as a being who directs or makes use of it by incorporating it into ensembles, or as a being who serves it by providing materials and elements' (Simondon, 2001: 119–20).

In the first chapter I looked at the differences between machine and human memory, which Simondon argued undermined the cybernetic analogical identification of the two. Where the machine can retain 'very complex, detailed, precise monomorphic documents' (Simondon, 2012: 75), humans excel at the memory of patterns, forms and meanings. It is the synergetic conjunction of the operation of these two forms of memory that Simondon presents as an example of the kind of operational continuity, which he describes as 'profound and essential', although he is clear that it is not just the function of memory that need be involved: 'coupling occurs when a single and complete function is carried out by the two beings'. Once again we can witness here the importance Simondon places on the productive and functional resolution of a disparity.

Despite concerns regarding the cybernetics misidentification of organisms with machines this coupling is undoubtedly cybernetically inspired, not just because of the importance Simondon gives to informational regulation, but also the confluence of recursive causality and finality: 'In self-regulated functioning, all causality has a sense of finality, and all finality has a sense of causality' (Simondon, 2001: 119).

The presence of such recursive causality should not be understood as occurring only with advanced informational technology. Such human-technology causal recursion is present in the use of even the most basic tools such as planing a piece of wood, in which the craftsperson can gauge the resistance of the grain of a piece of wood from the feedback he or she feels through the tool. Alienation describes a poverty of the integration of psychosomatic aspects of the person (for example, affect, emotion, perception) in the recursive operational continuity of the human-machine relation. According to this analysis both the owners of technology and the labour force are equally alienated for neither maintains such a relationship of continuity with technical individuals. The problem for each of these is different and rests on a basic misunderstanding of the informational aspects of modern technological individuals (such as control mechanisms and indeterminacy) and ensembles, which puts both out of step with technological reality:

> Labour has an understanding of elements, and capital has an understanding of ensembles; but bringing together the understanding of elements and the understanding of ensembles does not create an understanding of the *intermediary and unmixed* being which is the technical individual. Element, individual and ensemble follow one another on a temporal line; the man-of-elements slow in relation to the individual; but the man-of-ensembles who has not understood the individual is not ahead in relation to the individual. (Simondon, 2001: 118)

As such neither a proposed collectivization of the means of production (such as proposed by Lenin, for example) nor a Feenbergian humanism offers a solution to this kind of alienation as they don't address a transformation of the human-technology coupling at a level that would transform the quality of its operation for either partner. Such is Simondon's concern for the quality of the human-machine coupling that he also understands the machine as impoverished (Simondon calls it a form of enslavement) if its relationship with the human is weak in this sense.

NATURE

Simondon was swift to recognize the significance of the informational aspect of modern technology and its importance for human-technology relations as well as for a 'technological philosophy'. That cultural understanding remained out of step with these developments was extremely damaging, given it meant that technology continued to be understood in the nineteenth-century technocratic mode as 'a philosophy of human power through technics' (Simondon, 2001: 126), a human power that also led to an unbalanced relationship with the natural world.

At times Simondon comes close to a Heideggerian critique of technocratism both for its will-to-power and for its destructive attitude towards nature. It is precisely the failure to understand the informational aspect of technology (that it is open to both indeterminacy and regulation in ensembles) that results in a technocratism that is autocratic in that it merely sees the relation between technology and nature through the lens of domination and enslavement. This misunderstanding is compounded by maintaining an understanding of technology through the schema of thermodynamic heat engines, which sees it as substantial and operationally deterministic. As such Simondon fully understands why some substantive theorists of technology are threatened by it and fully acknowledges its 'enslaving violence':

> To this phase [industrial thermodynamic] corresponds the dramatic and impassioned idea of progress as the rape of nature, the conquest of the world, the exploitation of energies. The will for power is expressed in the technicist and technocratic excessiveness of the thermodynamic era, which has taken a direction both prophetic and cataclysmal. (Simondon, 1980: 8)

We find here, then, a tension between Simondon's positive genetic account of progress in which technology unveils and makes use of previously hidden potentials and this description of a technocratic cataclysm that is 'guilty of a violation of the sacred' (Simondon, 2001: 127). We have already seen that Simondon has described a sense of the sacred as those key points that give a reticulated structure to the magical mode of human-nature relation. As such technocratism comprises a destruction of this mode by shifting to the extreme technological pole of the initial religion-technology phase-shift. Such technocratism needs to be tempered by an aestheticism that recognizes 'natural integrity':

> To build a bridge over an arm of the sea, to link an island to the mainland, to cut through an isthmus, is to change the configuration of the earth, to violate its natural integrity. (Simondon, 2001: 127)

It is the notion of integrity in this passage that is indicative of the regulative proposal Simondon will utilize to resolve this tension. So although Dumouchel's assertion that 'there is no "nature" in the sense of events and processes which are essentially different from those which are produced artificially' (Dumouchel, 1995: 268) does describe the mode of genetic development that is technological, it fails to take into account an aesthetic and regulative dimension that the sacredness associated with 'natural integrity' implies. It is not the whole story. It is this respectfulness to the natural that prevents Simondon's account of technical invention from being unconstrained by ethical or aesthetic values.

To put it another way, Dumouchel asserts the progressive transductive individuation of technology without reference to the other modes of amplification Simondon proposes, namely modulation and organization. Just focusing on the transductive operation is to miss the importance of structure for allagmatics.

In attempting to tie together and resolve all the issues discussed so far I will now turn to describe the role that culture plays in resolving some of the tensions in Simondon's account.

NOTES

1. Ellul's argument that technicity entails the removal of indeterminacy from systems corresponds to contemporary claims made for Big Data. Such a position logically leads to hypertelia, as I'll argue in chapter 8.
2. We will discuss Feenberg's constructivism in more detail below.
3. For a definition and discussion of transduction, see chapter 2.
4. Simondon's entire project is underpinned by the aim to forge a new genetic encyclopaedism that takes as its goal the thinking of the genesis of *all* things. It should by now be clear that his work on the different regimes of individuation is central to this encyclopaedic project. As we will see, his work on technology connects this project to a new description of humanism in that its aim is specifically to combat a certain form of alienation.
5. The description of the reticular milieu corresponds with that given in the third step of the image cycle in *Imagination et invention*.
6. Ideology will be discussed further in chapter 8.

Chapter Six

Culture and Technology

So far in my analysis of Simondon's account of technology I have described the technological mode of individuation as well as some of the problematic relations between humans and technology (alienation) and nature and technology (technocratic domination). Ultimately the resolution for these problems will be found in the more fundamental relationship of culture and technology.

It is worth recalling that these problems are specific to industrialization and that prior to this period of technical development Simondon identifies a period of phasic harmony between technological development and culture. That culture has become out of phase with technics, which is the major theme of *The Mode of Existence of Technical Objects*, does not entail that it must remain so. That is to say that, unlike Heidegger and Ellul, Simondon does not see the situation as irretrievable, and unlike Feenberg he sees the application of humanistic values to technics as the cause of and not the solution to the problem.

To understand the meaning of Simondon's demand for a technical culture as a solution to these problems we need to be clearer about the dynamic relation of culture with technics, specifically that the problem of culture is actually inherent to technics itself. A useful place to start with clarifying this is Simondon's article 'Culture and Technics' (1965), which outlines the ambitious scope of Simondon's philosophy of technology.

For Simondon culture is the domain concerned with values, and as such, its response to industrialization is as a defence mechanism. However, Simondon reminds us of the agricultural origin of the meaning of the term *culture*, which is founded on a set of techniques that involve either working directly on an organism itself (for example, husbandry, grafting) or working on the organism's environment in order to force adaptation (for example, cultivation). Techniques in agriculture are thus utilized to intervene in an organism's individuation. As mentioned in the previous chapter, changes brought about by working directly on an organism often lead to instances of artificiality, dysfunction and hypertelia in organisms. As such Simondon is explicit that culture is the use of technique for cultivating the human species. In a striking passage that resonates with contemporary interests regarding posthumanism, Simondon states that

> whether or not he wills it, man is the technician of the human species; a form of feedback loop operates in human groups, alternately comparable to either the farmer or cultivator who prepares the soil, or to the gardener or breeder who deforms species and obtains new varieties. When the feedback loop is comparable to the cultivator who acts on the soil rather than the plant, we speak of technique: man acts on the environment he exploits, transforms and develops, and in this case man only acts on himself indirectly by means of environmental potentials. By contrast, the contemporary usage of the term 'culture' is paradoxical: the word is employed to designate the result of direct action of man upon man, comparable to that of the gardener or breeder; it remains a question of techniques, techniques for constituting collective or individual habits, or training in the various prohibitions and choices that define a psychosocial personality. (Simondon, 2015:18)

In this passage we find a number of significant ideas that help draw together several of the concerns that have already emerged in our discussion. The inevitability of the situation that the human species is responsible for its own evolution and that this development is guided by techniques that presumably, given the analogy with agriculture, could also lead to hypertelic dysfunction is striking. What is clear is that the opposition between culture and technics is to a certain extent a false one. Instead it must be understood that culture itself is composed of techniques and that where the conflict occurs is between the kind and scope of the techniques of culture with those of technics.

An important distinction Simondon makes is between those techniques that he describes as *intra-cultural* and those that are described as *technics* more properly. For Simondon culture is the set of techniques and corresponding values that operate to maintain stability at the order of magnitude of the human group. As such culture is something that is passed down generationally and corresponds to the cultivation performed by the breeder, which is directly on people. Simondon is here thinking of techniques that are local in the magnitude of their operation, such as the use of language, institutions and customs or the use of 'closed' pre-industrial technology such as water pumps, which may be designed differently in different groups but all serve a single dedicated function whose operation tends not to overflow the bounds of the group. At this level technology is regarded as functionally fulfilling predicated ends that serve the stability of the bounded human group.

In contrast to these *intra-cultural* techniques Simondon describes the *schemes of technicity*, which describe technologies whose operation tends to work indirectly on man via feedback effects through changing his environment, provoking adaptive and collective transformation. Technicity at this scale, and Simondon has in mind industrial technology here, overflows the bounds of the intra-cultural both in its environmental impact and in the manner in which it spreads across the planet impervious to cultural resistance due to the adaptive changes it necessitates. As we have seen, the mode of individuation of technicity not only engages environmental potentialities but also creates new ones, a tendency that means it is more inventive towards the future and so generally less conservative than culture. This inventiveness is expressed not just in the way technology overcomes the restrictions of the physical bounds of the cultural group but also in that it exceeds the utilitarian and functional view culture has of it, and is creative of new desires that are productive of further future technical developments.

What is at stake in the difference between the *intra-cultural, minor* techniques and the *pure* and *major* techniques, as Simondon also refers to them, goes beyond a choice between the destruction of cultural values or the enslavement of technology and concerns the evolution of the human race itself.

The opposition of culture and technics can now be understood as a problem that is internal to technics itself in that minor and major techniques have fallen out of phase with one another. However, this dephasing should not be viewed as cataclysmal but as an opportunity with evolutive capacity.

Given this cosmic vision Simondon is defensive of major technical gestures such as programmes to explore space, which mark the extreme boundaries of current technical achievement, not because of any utilitarian value but because they broaden the evolutionary capacity of the species rather than just settling for what already exists. Such technical achievements are significant as they transform the human species' relation with its environment and are thus responsible for the cultivation of the species. It is because of this Simondon describes such ambitious technical undertakings as 'great autonormative gestures' because through transforming this relationship they are also productive of new values, norms and forms of culture.

Given such enthusiasm for great technological gestures and the seemingly unbridled expansion of technicity it is easy to see why Dumouchel understands Simondon as arguing that there is no technology 'which can respect what is', given it is continually productive of new potentials, ways of thinking and values. However, to get too enthused with this wholly cyberpositive, runaway conception of technical development is to betray the role of structure and regulation that is also important for Simondon. To ignore this aspect of Simondon's account would make him little different to Ellul in that the broadening role of technicity would lead to the eclipse of the human.

Simondon is interested in developing a new humanism and although he shared many of the same concerns as Heidegger, Ellul and Marx, such as alienation and the despoliation of the natural world, he also suggested very different solutions that didn't require the hylemorphic application of humanist values upon technological use and development. Instead Simondon argues for the requirement of regulatory and ecumenical gestures at different levels of individuation, enabling a suitable relation between culture and technicity to develop. In the account of the relation of culture and technicity outlined above Simondon describes man as the 'technician of the human species', a description that designates to the technician an important role for developing this relation.

THE ROLE OF THE TECHNICIAN

Where it is the artisan who represents the appropriate and complementary relationship between the human and the use of tools, it is the figure of the technician who is the exemplar of overcoming human-technology alienation in the industrial era. The technician exists 'at the same level' as the technological ensemble and can thus be integral in its regulation and coupled with it in such a way that the human-machine dyad forms a functional and operational whole.

An important aspect of this coupling is that of meaning, which the technician introduces through invention. As a concretization of forms machines are not able to interpret functions but are the result of concretization according to schemata of operation. As such it is the technician who can link machines together through the analogical use of thought, which enables the human to think the technological in its genetic becoming, that is to say, as mechanological invention:

> To invent is to make one's thought work as a machine works, neither according to causality, which is too fragmentary, nor according to purpose, which is too unitary, but according to the dynamism of lived functioning, understood as a product, and understood also in its genesis. The machine is a being that works. Its mechanisms gives material expression to a coherent dynamism that once existed in thought, and that was thought. (Simondon, 2001: 138)

As well as being a process that involves the discovery and *actualization* of potentials, invention is simultaneously the stage of actualization described in the cycle of the image. In the cycle of the image there is a gradual concretization of images from that of the development of a world through to systemization using schemata. In technical invention there is a reciprocal relationship between imagination and technical objects in that the latter are crucial for developing schemata of understanding, which are necessary for resolving problems, the solution to which often leads to the invention of new technical objects. The cycle of image-objects is thus part of the same objective-individuating process as that of technical objects. As such, although invention requires the presence of a real problem, it also requires the application of the thought of the inventor to guide the problematic structure to its resolution. That the schema that thought uses is developed through its imbrication with technicity completes the loop.

It is in regulation and invention, understood as the development and maintenance of the concretized relations of humanity with technology, where Simondon finds the overcoming of alienation. This overcoming is made more likely with the development of informational, networked technologies. Whereas thermodynamic machines, due to their relative closure, are more limited in the amount of information they produce that is available for interpretation, the development of reticulated ensembles requires a greater role for technical interpretation to ensure regulation. What's more, any increase in information from machines also increases the amount of possible indeterminacy in operation and thus the potential for further invention.

This indeterminacy in the machine's operation is what Simondon refers to as its 'openness' and is precisely what he feels contemporary culture misunderstands about it, instead still understanding technology as closed and substantial, like in its intra-cultural or minor stage. This attitude is one that Simondon deems unfair and that he likens to taking a reductive attitude to a painting such that one only understands it as 'an expanse of dry, cracked paint on a stretched canvas' (Simondon, 2001: 146).

It is this recognition of the openness of networked technology that makes Simondon's analysis of technology so relevant for thinking about our contemporary media technology situation. This openness of the milieus of information technologies means they have extensive potentials for forming a broad number of unities. This has helped establish them as environmental. This transition has also led to a shift in the conception of labour from being hylemorphic to being a process of modulation. The openness of technological operation means the inventive powers of the technician, as a new kind of worker engaged in technical activity, is required to forge and regulate these new unities of meaningful operation.

> Man understands machines; he has a function to play between the machines rather than above the machines, so that there can be a genuine technical ensemble. It is man who discovers meanings (les significations): meaning is the sense that an event takes in relation to forms that already exist; meaning is what makes an event have information value. (Simondon, 2001:138)

It's true that writing when he was, Simondon perhaps didn't foresee the extent to which the open nature of our technologies would truly transform them into being ubiquitous in human environments, blending mobility with extensive infrastructures, computation with Big Data, and even operating independently and beyond the threshold of our perception and cognition.[1]

The requirement to instigate a culture with an adequate appreciation and understanding of contemporary machines leads Simondon to propose the need for mechanologists who would hold positions corresponding to sociologists or psychologists of machines in that their role would be to ensure the cultural understanding and significance of technical realities.[2]

We can also discern here a crucial difference that separates Simondon's conclusions regarding technology from Heidegger's and Ellul's. Where these two philosophers understand the dangers of contemporary cybernetic technology as either a removal of the human with all its fallibility from the system to achieve a perfection of automation or the diminishing of the human as 'standing reserve' for systemic purposes, Simondon sees in the informational nature of cybernetic technology an indeterminism that is its saving grace. Unlike with the more determined nature of thermodynamic technology he perceives the possibilities opened up by the greater indeterminacy of informational technology. This is a profound contrast to the reduction of humanity to the status of machine found in Wiener's work[3] and instead Simondon argues that the human has an important place *among* the machines, working with this openness.

For Heidegger the alienation we experience in relation to modern technology is from lacking an understanding of the essence of technology and thus being in an unfree relation to it. Technology then comes to have a conditioning effect on the human relation with other beings and Being in the form of an enframing or unconcealment. For Simondon technology describes a revealing of a rather different kind to Heidegger's, the discovery and invention of new potentials for fresh individuations. Although, like Heidegger, Simondon is interested in how technology transforms humanity's relation to the world, his concern is with the operations by which such a transformation occurs, rather than one fixated on meaning. It's true that with his description of the damming of the Rhine Heidegger is concerned with the transformation of a natural phenomenon. This is not so much because of the disruption of 'natural integrity', however, but the transformation of the mode-of-revealing mankind has in relation to the world.

Despite the description that Simondon gives of the cataclysmic effects of thermodynamic technology for the natural environment, his final analysis is far from being as pessimistic as Heidegger's. Just as Simondon looks to informational technology to alleviate alienation he also claims that the development of technical ensembles governed by informational theory leads to a more stabilizing form of technology and must be recognized as having the potential for supporting a different relation between nature and technology:

> The machine, as an element in the technical ensemble, becomes the effective unit which augments the quantity of information, increases negentropy, and opposes the degradation of energy. The machine is a result of organization and information; it resembles life and cooperates with life in its opposition to disorder and to the levelling out of all things that tend to deprive the world of its powers of change. The machine is something which fights against the death of the universe; it slows down, as life does, the degradation of energy, and becomes a stabilizer of the world. Such a modification of the philosophic view of technical objects heralds the possibility of making the technical being part of culture. . . . Today, technicality tends to reside in ensembles. For this reason, it can become a foundation for culture, to which it will bring a unifying and stabilizing power, making culture respond to the reality which it expresses and which it governs. (Simondon, 1980: 9)

Here we can discern a rebuttal of Heidegger's antipathy to technological culture. Simondon not only points to the stabilizing effect technology can have but, further still, that technology in its capacity as 'natural object' is actually a tool to combat entropy (once again we can discern at play a neotenic effect here). What's more, he is also clear that what's required is that this new kind of technology can found a new kind of culture, that is to say new values and norms. The technician and inventor are thus the human avatars of the dual movement of regulation and invention necessary to maintain the metastable dynamic of this negentropic culture.

Also of importance is the reticulated nature of contemporary technical ensembles and how this is understood analogically with the magical mode of human-world relation. Following the initial dephasing of magical unity we saw that the human relation to its milieu splits into the phases of technicity and religion. Simondon also describes this split as that into figure and ground such that technics 'retains the figural characteristics of the primitive complex of man and the world, while religiosity retains the ground characteristics' (Simondon, 2001: 173), such as describing a relation with universality and totality (for instance via deontological ethics). It is this figural quality of technicity that describes its role in the dislocation of the reticulated key points of magical unity into 'detached' singular sites that have lost a close relationship with the ground. That they are also multipliable and plural also makes them particularly disruptive for unity.

Thus just as concretization marks a progress in the unity of functioning of a particular technical object, the overall progress of technological development is understood as a move away from overall unity. Technicity is concerned with operations that occur at a particular time and in a particular place such that 'adding technical objects one to another can neither re-make the world nor regain contact with the world in its unity, which was the aim of magical thinking' (Simondon, 2001: 174). Despite this Simondon does see networked informational technology as an opportunity to reverse this situation through the implementation of its openness. His argument for how this new form of technology can balance the religious and technical domains requires a closer look at the aesthetic relation.

THE ROLE OF THE AESTHETIC

Simondon is clear that the technocratic drive that has made industrial technology so alienating is at the extreme technical pole of the technology-religion relation. It marks a position concerned only with the technical and figural and is therefore out of balance with both the universal attitude of religion as well as any ground. It is for this reason that the technocratic attitude, which reaches its peak with the development of thermodynamic technical individuals in the twentieth century, is viewed as so cataclysmic, for such is its concentration on exploiting the potentials of the natural environment it neglects the quality of its aesthetic insertion into that environment. It is therefore also particularly out of balance with other kinds of relation to the world.

For Simondon aesthetics is not primarily concerned with representation. For him all objects have aesthetic qualities the extent of which depends on the manner with which they are embedded in the world, something he calls 'aesthetic impression'. This impression

> implies feeling the complete perfection of an act, a perfection that objectively gives it a radiance and an authority by which it becomes a noteworthy point of lived reality, a node of experienced reality. The act is a remarkable point of the network of human life inserted in the world; from this remarkable point to others a superior kinship is created which reconstitutes an analogue of the magical network of the universe. (Simondon, 2001: 180)

The reticulation of acts of aesthetic impression thus constitutes an analogue of the networked structure of key points in the magical universe, although not at the level of magical unity but as situated in the relation between the subjective (religious) and objective (technical) phases. Thus for Simondon an aesthetic object straddles these two domains, enabling a balance in which the objective use of a technique (for example, building churches/shrines etc) can engage the subject with universal qualities (for example, a religious attitude) while remaining immanent in respect to individuation. The aesthetic enables a reconnection of figure with ground, transcendence with immanence, although not through the dissolution of the subject-object division; hence the analogous quality of the network of aesthetic impressions to that of magical unity. This connection of the subjective and objective also recalls the importance of the relation of act with emotion that is the very definition of the transindividual. With this aesthetic appreciation of technicity we also discern an approach towards an ethico-aesthetic theory of value. In a letter to Jacques Derrida Simondon outlines the importance of what he calls techno-aesthetics:

> The techno-aesthetic feeling seems to be a category that is more primitive than the aesthetic feeling alone, or than the technical aspect considered from the angle of functionality alone (which is an impoverishing perspective). (Simondon, 2012: 6)

The primitiveness of the techno-aesthetic originates from how it intersects the relations between the subjective and objective, the figural and ground, the singular and the universal, the technical and the natural, in what Simondon calls 'an intercategorial axiology' (Simondon, 2012: 2). In the letter Simondon provides a range of examples of the unifying nature of the techno-aesthetic in the relation of technicity with both the human and natural world. Of the Garabit viaduct, he writes that

> it's beautiful also because it's in the middle of nature. The viaduct traverses nature and is traversed by it. . . . This is an example of a techno-aesthetic work: perfectly functional, successful, and beautiful. It's technical and aesthetic at the same time: aesthetic because it's technical, and technical because it's aesthetic. There is intercategorial fusion. (Simondon, 2012: 2)

However, it is not just the functionality that is aesthetic here but also the way that the viaduct's structure complements the environment in which it is placed such that there is an aesthetic concretization. It is a matter of how the technical object complements that which is already there so that the coupling not only works at a technical and functional level but also at an environmental level.

This approach, which is sympathetic to technical operation and its complementary integration into a natural environment, thus provides a way to contrast Simondon's judgement on technical structures to Heidegger's. Both thinkers ultimately reject an unbridled technicity for similar reasons: Simondon's account of cataclysm is not too far from Heidegger's 'standing reserve'. There is undoubtedly also a similarity between the two thinkers in that both offer a critique of technicity concerning the interweaving of the aesthetic with functionality. In Heidegger's example of the windmill we see a similarity to Simondon's thought that an 'acceptable' technological object should complement the natural phenomenon it utilizes. The difference between the two thinkers is more apparent when we consider Heidegger's horror at cases in which the 'essence' of the natural phenomenon is sullied by technology such as when the Rhine's mythic symbolism is destroyed by a dam that reduces it to just a functional part of a hydroelectric project. Where Heidegger signals a refusal here, Simondon offers a subtler response, because for him the techno-aesthetic itself is situated so closely to both the magical and religious. Because of this the aesthetic is not primarily about contemplation, which is ultimately Heidegger's retreat position, but rather is concerned with operation via action; this leads him to a very different understanding.

> Contemplation is not techno-aesthetics' primary category. It's in usage, in action, that it becomes something orgasmic, a tactile means and motor of stimulation. When a nut that is stuck becomes unstuck, one experiences a motoric pleasure, a certain instrumentalized joy, a communication—mediated by the tool—with the thing on which the tool is working. (Simondon, 2012: 3)

Simondon's judgement of a dam would not be made just on the fact that it blocks a river but on *how* it did so, on the quality of the integration of the dam's structure with the operation of both dam and river. For it is the coupling of dam and river that would concern Simondon both at the level of pluri-functionality[4] and also with how the technical object intervenes in the environment qua environment; that is, not only complements that which is already there, as we saw with the Garabit viaduct, but makes of that place a key point where it 'completes and expresses the world' (Simondon, 2001: 185).

Technical objects for Simondon are 'mediators between nature and the human' (Simondon, 2001: 9). As such even the smallest act has aesthetic meaning, as the example of the loosening of a tightened nut demonstrates. The fundamentality of the techno-aesthetic, that it occurs at such a primitive level that it doesn't require contemplation but 'connects with the elaboration and satisfaction of biological desires themselves' (Simondon, 2010a: 232), is truly where Simondon diverges from Heidegger.

> The aesthèsis, the fundamental perceptive intuition, is part of a culture. It acts like a pre-selector, separating the acceptable from the unacceptable, and determining whether one will accept or refuse. (Simondon, 2012: 4)

His proposal for the overcoming of alienation is also directly related to the techno-aesthetic. Just as the aesthetic pleasure of using basic tools such as wrenches or drills relies on the operation of a recursive causality between human and world, mediated by the tool, so Simondon's proposal for the alleviation of alienation demands the technical object remain a mediator for human-world informational recursion. That is to say, it remains *meaningful* in that the mode of affective engagement with which the human engages is comprehensive, rather than narrow such as on many production lines.[5] Thus Simondon can say of the aesthetic object that 'it is never, strictly speaking, the object that is beautiful: it is the encounter, operating by way of the object, between a real aspect of the world and a human gesture' (Simondon, 2001: 191).

As an encounter the beautiful is always found in the process of a mediated impression, thus artwork and technical objects become sites that stimulate such encounters. The aesthetic encounter always concerns an operation that mediates between the singular act and a higher level of unity, it is at once situated between that which is particular and that which would be universal. As such Simondon's aesthetic always involves a unifying movement, whether that of the pluri-functional and concrete or of the intercategoriality that occurs in a technical act.

In his later work Simondon was impressed by the development of technical networks whose structure he considered as isomorphic to magical unity in that they were reticular and included key nodes, which held influence that could be amplified throughout the whole network. In this way Simondon extends his techno-aesthetic to include not just how singular technical objects might balance intercategorically but also how networks of technical objects, operating using informational circuits, enable a harmonization throughout the technical and psychosocial domains.

The aesthetic tendency is 'the ecumenism of thought' in that it always tends towards unification, it is a synthesizing operation by which the development of a reticulated structure in one domain aligns with that developed in another. It is perhaps unsurprising then that Simondon recognized in the literal reticular structure of informational ensembles a resemblance with the reticulated structure of magical unity to which aesthetic thought asymptotically aspires.[6]

EDUCATION

Another important suggestion for helping to regulate the relation between technology and culture is via education. Given the importance of technology for the contemporary world it is clear that not all social problems can be addressed in purely cultural terms. What is required for the solution of many problems facing humanity is an understanding of the relation between man and the environment, which is predominantly mediated through technology.

Simondon's concern is that there is an over-emphasis given to teaching traditional cultural content during the formative years of education that 'fixes norms and basic cognitive schemas . . . according to cultural contents inherited from the past' (Simondon, 2015: 21). Although this education is 'affecto-emotive on the one hand, and perceptuo-cognitive on the other' (Simondon, 2015: 21), it importantly neglects those affective and technical schemas related to technics.

It is perhaps little wonder that those who have undergone a traditional cultural education, which also tends to be non-technical, are resistant to or, at the very least, ambivalent to matters technological, when the technical schemas required for understanding it and the affective modes it enables don't form part of their worldview. Simondon therefore proposes a pedagogical shift such that technicity is taught simultaneously with cultural education, thus enabling the student to achieve a unified comprehension of the world. This would enable students to learn not just technical schemas but also how values individuate and are not cultural constants to be imposed. The enrichment of the imagination with technical schemas would also strengthen the solving of problems by providing the ability to select whether a problem should be analysed as purely cultural, technical or mixed.

ETHICS

Having now described the major components of Simondon's philosophy of technology in this conclusion I want to explore the tension that holds between the two movements described in relation to a Simondonian ethical theory.

On the one hand there is the productive account Simondon gives of the mode of existence of technical objects as progressive invention through the ongoing concretization of potentials and virtuality. From this perspective we saw how Simondon can be understood (as I have argued Dumouchel does) as arguing for the impossibility of holding a normative position regarding technology because the progressive invention that technicity describes cancels such possibility.

On the other hand there is an account that focuses on regulation, both in the special place reserved for the notion of magical unity as indicative of a relation to the world that coheres between transcendence and immanence and in the account of techno-aesthetics and the proposed solution for alienation through the establishment of a specific mode of human-machine coupling. This aspect of Simondon's philosophy also underpins his genetic encyclopaedism, one of whose aims is to help keep in balance the various cultural phases described in 'The Limits of Human Progress'.

The position that is reached in order to resolve the tension between invention and regulation is that the ontogenetic individuation of technicity entails the requirement for a corresponding ontogenetic intercategorial axiology. This perspective can also be discerned in the core argument of *The Mode of Existence of Technical Objects*, which is that culture is lagging behind in the individuation of values appropriate to the current stage of technicity and thereby responsible for a new kind of alienation.

What I wish to address is: Does Simondon describe such an ethical theory by which he can resolve the tension in this dynamic? Is the importance of invention in Simondon's ontology really inconsistent with the significance he gives to regulation? Interestingly this question mirrors Kant's problem in the third critique, which was of how to explain the apparently teleological nature of the organism in a mechanistic universe. My answer to this will be that this tension can be resolved by being clear that when Simondon discusses regulation through recursive causality he is not mistaken as proposing a teleological explanation.

To begin addressing this I first want to examine what an ethics based purely on becoming would look like for Simondon. The account Simondon describes is not one of fixed norms (such as we find with Feenberg, for instance) but in norms which themselves individuate 'under the pressure of becoming' (Combes, 2013: 64). It should come as no surprise that Simondon's ethical theory is described in allagmatic terms as one of individuation. That is to say that ethical development occurs via processes in which metastable normative structures transition through ongoing ontogenetic operations involving acts and values. Norms can be understood structurally in that their individuation brings coherence to a collective system, that is, they help maintain the compatibility of individuals with one another and with their environment.

Although norms may appear stable they are in fact metastable, that is to say that they can be disrupted either from within, by social disindividuation, or by a singularity encroaching from the exterior. For example, the invention of a technology can bring with it its own demands and values that contrast with established norms, creating a disparity whose resolution may well require the transition to a new system of norms, which enables a compatible structuration of the social in line with the new technical situation.

Values enable the transductive development of norms, that is, their conversion from one normative system into another. Although I've mentioned technicity as a source of values, Simondon is clear that individual acts are also a source of value and hence normative change. An individual who acts due to a problematic situation in relation to its milieu is acting ethically and that act is the source of values that can challenge the established norms of the system. A single act on its own will likely not transform norms but when that act is amplified in a network of acts it can transductively overwhelm a metastable situation and lead to a fresh transindividual individuation. Rosa Parks's individual act of resistance in the face of a normative situation is an example of an act leading to the transformation of a saturated situation via a network of associated acts, linked and transductively amplified. Similarly, Simondon asserts that the inventions of technicians can also have a transformative effect through the symbolization and signification these attain throughout a collective, or via requirements of their operation.

The mode of individuation of norms is through an 'amplifying transfer', which describes the process by which acts and values relate to one another in a reticulated process of resonance. We can identify a politics nascent in this ethical account that, for a political transformation to be efficacious, requires it being able to initiate a transductive contagion of values/acts across a collective domain, thus instigating social transformation via restructuration. Importantly such a theory of ethics proceeds clearly as an operation of physical individuation. As Combes writes, 'We can no longer distinguish between the level of sense or meaning and that of *physis*' (Combes, 2013: 64). As such Toscano (2012) correctly discerns the political theory implicit in Simondon's work as an energetics rather than a dialectics:

> The pre-revolutionary state is the 'very type', according to Simondon, of the psychosocial state which a political science of metastability should concern itself with 'a state of supersaturation. . . . Where an event is very ready to occur, where a structure is very ready to emerge. (Toscano, 2012: 92)

Given the priority that Simondon assigns operation over structure it shouldn't be a surprise that it is the act that is the focus of Simondon's ethics. The ethical act of the individual is a concretizing act, as it were, in which the individual regulates its own actions in relation to its milieu. As such it is an ethics of auto-normativity, which is in direct contrast to a universalized ethics such as Kant's, which begins with the substantiality of the actor and so is described by Simondon as a substantialist ethics. Contra such substantialism, there is no fixed goal or end in Simondon's ethics, for this would be the hylemorphic imposition of an eternal form or *telos* onto individuating being. Instead the sense of ethics must be grasped from within the allagmatic disindividuation and individuation of norms itself, understood as metastable structures that are both regulating during the period in which they provide compatibility, but also open to further individuation when crossed and overflowed by disparate values.

Moreover, we may go further with the operative nature of this system by also drawing on Simondon's theories of the image-cycle and the transindividual. The transindividual relation is one of collective resonance. As we saw, this relation can result in a more or less coherent social structuration (in-groups and out-groups). I also described in that chapter the important role of the image for invention, which involves the externalization of the image as image-object. As such the image-object is always already imbued with values and meaning as well as having causal power in structuring the transindividual relation through its role in transforming humanity's relationship with nature:

> The circular causality that runs from the mental to the objective real through social processes of cumulative causation also runs from the objective real to the mental. Every image is susceptible to incorporation in a process of materialising or idealising recurrence. (Simondon, 2008: 13)

The mode of operation by which ethical norms individuate is one that must involve this circular causality that flows between image and image-object, which it must be remembered is a process that involves the psychic at all of its stages (for example, affect, sensation, emotion, conceptual) as well as transindividuality. As such we can understand Simondon's project truly as a reformed socio-cybernetics wholly founded on his reformed notion of information operating across all regimes of individuation. Crucially, such operation is always differentiated, that is, it always operates topologically such that resonance 'phenomena' entail an intercategorial axiology.

Ethics then must be a process of collective transformation and Simondon understands this as meaning both that norms develop through a process of amplificatory transfer and that the reticular nature of newer technologies (including media technologies) can change the nature of this amplification and thus entail a transformation of values.

Where does this leave the regulatory side of Simondon's account? Simondon's focus on the operation and development of recurrence at all levels requires a consideration of the regulatory nature of this operation. There is a straightforward regulatory operation of reticular structure, as Combes succinctly points out, regarding participation:

> Although we may change tools or construct a tool ourselves, 'we cannot change networks or construct a network ourselves' (Simondon, 2001: 221). (Combes, 2013: 67)

Given, no doubt to the disappointment of many technophobes, that participation isn't always a choice we are at liberty to make, it must be made clear that Simondon is no techno-determinist despite the constant account I've given of culture lagging behind technology or technology providing values leading to normative transformation.

What needs to be reiterated is that at several places throughout his work Simondon highlights the special place of reflexive philosophical thought for regulating the relation of technicity with the social. In particular it is used for both tracing the individuation of being and understanding the regulatory aspects of unities that are formed. Throughout his work Simondon consistently accounts for phenomenon at all levels of reality in terms of a relation of openness and closure (for example, transduction/modulation, operation/structure, value/norm, liberation/alienation) and that in this chapter has involved the relation of the closure of culture to the openness of technical individuation (although this relation can be understood as being internal to technicity itself).

Although it is not possible to derive a clear political project from Simondon's work, deliberately so as to have done so would be to assert a fixed hylemorphic social structure, it is possible to point to places in his work where he proposes the requirement for reflexive thought in order to evaluate and guide the *functioning* of the social, particularly in respect to the individuation of technics. In 'Culture and Technics' Simondon is at his clearest with respect to the role of reflexive thought and philosophy, which is to be able to identify the position of problems and how best to address them after they arise from the ongoing oscillation between culture and technicity.

> Culture and technics cannot complement one another while remaining in a static position; they can become complementary only through a kinematic process of oscillation and inversion, according to a regime whose ability to grasp each problem is perhaps the highest task philosophy can assume. (Simondon, 2015: 23)

What this practice involves is utilizing a truly interdisciplinary approach to identifying and solving social problems. This can't be achieved only through an understanding of traditional culture, such as via symbolism, but also requires the application of technical schemas of understanding. Hence why Simondon's cybernetically inspired encyclopaedism is of such importance.

It is also apparent that the notion of feedback, also derived from cybernetics, is crucial to this approach. Simultaneous to the operation of technicity on the social, is the operation of the social on technicity, predominantly through the regulatory dimension of culture. Simondon's aim for philosophy is that it can regulate culture in such a way that the latter better understands and values the operation of technics. An understanding and integration of technical schemas into culture can only aid social regulation, given that it will make possible the resolution of properly technical problems using a suitable 'mode of analysis' (Simondon, 2015: 23). As such reflexion must not limit itself solely to theoretical excursions but must include acts, for as we saw in his ethics, acts are also a source of values, both aesthetic and ethical. Indeed, to return to the key regulatory concept from *The Mode of Existence of Technical Objects*, 'the act is a remarkable point of the network of human life inserted in the world', which along with informational technologies 'reconstitutes an analogue of the magical network of the universe' (Simondon, 2001: 180).

The importance of the concept of the unified magical mode of being in the world underlines the importance Simondon places on balancing the religious mode of relation to the world with that of technicity. The neglect of the religious mode of being is something that Simondon felt was dangerous and could lead to technocratic thinking. Hence in his letter to Derrida regarding the foundation of a College of Philosophy Simondon is clear about his displeasure at the lack of religious content:

> If our fundamental aim is to revitalize contemporary philosophy, we should first of all think of interfaces, and nothing should be excluded *a priori*. There

> is no reference to religious thought and practice in your project. Why? (Simondon, 2012: 1)

This concern for religion is not due to some underlying faith. Instead it is because of the vital importance Simondon places on maintaining a balance between the universal ground (ubiquity and eternity)[7] and the figural and singular—that is, the transcendental with the empirical. His letter also echoes a passage in the third part of his thesis in which he bemoans the dominance of science over the affective-emotional in contemporary culture (Simondon, 2013: 249), which has led to a split between act and emotion. Such a split is entirely unhealthy from Simondon's perspective, given that transindividuation requires the combination of act with emotion. The revitalization of philosophy requires addressing the alienation of its time and finding a way to unify the technical outlook with the religious to rectify an excessive technicism. This is not a plea to attempt a unification of theology with naturalism, as each of these is mistaken in its understanding of transcendence and immanence. It is rather the need to unify the empirical and transcendental perspectives within the immanence of individuation.

It is with the reticulated structure of informational technological ensembles that Simondon identifies the creation of a situation analogous to that of magical unity within individuating being. That is, with the development of networked technical individuals into ensembles that are informationally connected Simondon identifies a new non-alienated role for humanity among the machines. According to Simondon, only man is able to informationally connect machines into true ensembles given only he can *make sense of* information (we must bear in mind Simondon's reformulation of the concept of information here). Machines work with forms and not information, that is, they lack the ability for invention and making sense of openness and indeterminacy. As such the new role for man is as mediator:

> It is no longer a universalising liberation that man needs, but mediation. The new magic will not be discovered in the direct radiance of the individual power to act, assured by the knowledge that gives the action effective certainty, but in the rationalisation of the forces that situate man by giving him a meaning in a human and natural ensemble. (Simondon, 2001: 103)

The networked structure of technical ensembles once more attaches the machine to the earth but this time in a unifying structure in which humanity plays a significant role. There is a balance here between the universality typical of religious thinking with the singular nature of technics. The network is both universal and singular. It is here that we find the motivation for Simondon's cybernetically inspired revitalization of humanism. As he states, 'every age needs to discover its own humanism' in response to the alienation it faces. The answer to the current state of alienation is the requirement to grasp the changed nature of technology and 'human reality'. This entails a new understanding of teleology and regulation. No longer is purpose something that is imposed on mankind from without but is now something that emerges immanently from the technical work that is undertaken in networks that humanity organizes:

> Cybernetics, a theory of information and consequently also a theory of structures and finalised dynamisms, liberates man from the constraining closure of the organization by making him capable of judging this organization, instead of suffering it by venerating and respecting it because he is incapable of thinking or constituting it. Man exceeds subservience by consciously organising finality, as he dominated in the 18th century the wretched necessity of work by rationalising it instead of suffering with resignation for making the work effective. (Simondon, 2001: 103)

The role of philosophical thought, cybernetically invigorated, becomes the invention and assessment of operational unities across varied and interconnected domains.

Simondon's techno-aesthetics and its focus on 'intercategorial fusion', then, far from involving a tension with the productive aspect of technics, complements it. If operation is what is primary, then our ethics and aesthetics should be concerned with the quality of the reality of this operation, and if operational realties also involve self-regulatory functions, then these should be judged as such. As Simondon asserts, all such self-regulatory functioning involves causality that 'has a sense of finality' but this 'sense' does not make it teleological, where that is understood as determining. This would be to make such power hypertelic and normative and to forget the ontogenetic process of individuation from which it emerged and from which it can be disindividuated. Simondon's theory of individuation is inventive and as we've seen it is inventive of realities at new levels, which operate as unities and thus can be understood as involving finalities. What Simondon makes clear is that such finality emerges from recurrent causality. It is real but it is neither necessary nor determining.

The ends are not to be imposed on the means, such as in productivity (capitalism) or communitarianism (communism). The system is not one that can operate homeostatically (Wiener), but is metastable and prone to change, if only that of inevitable entropy.

To return to Heidegger once again, it is clear that his and Simondon's positions are actually not too dissimilar even if their conclusions are very different. Both are in agreement regarding the dangers of industrial technology, and both agree that there is an epistemic component to this danger, for technology concerns a mode of revealing of the world. For Simondon, through invention technology reveals and indeed invents the modalities by which the world operates. It thus furnishes us with schemas for understanding. In contrast, and being deliberately blunt for the sake of simplicity, Heidegger takes the technical schemas of thermodynamic and cybernetic technology and considers them as substantial for technology as a whole in the mode of enframing. In a sense he fails to get past these schemas thereby forcing a retreat to an aestheticism rooted in art and poetry. What Simondon gives us is the understanding that not only are technical schemas always developing, but that we also shouldn't apply them to inappropriate domains in poor analogies. So although Simondon also values aesthetics, it's not as a retreat but as a mode of regulation of both the practical application of technology as well as our theoretical understanding of it.

Thus from a Simondonian perspective, Heidegger's understanding is limited as it emphasizes two particular technological schemas that are by no means the end points of a certain branch of epistemology. As Simondon's reformulation of the notion of information makes clear it is possible to go beyond these schemas and think and invent technology that is productive of completely different forms of relation of the human with the world, that is, as a coherent dynamism that sits, as I have argued causality must, between total indeterminism and determinism. This relation is also productive of ethical, aesthetic and epistemological developments.

Hopefully I have demonstrated that the inventive and regulatory aspects of Simondon's ontogenetic account must not necessarily result in an irresolvable tension. One could describe the resolution as that of regulation being the result of invention. Invention and regulation are in the end co-constitutive just as operation and structure are in Simondon's allagmatics.

Additionally, we have now established how it is that technology holds such a unique position within Simondon's ontology as it is involved in causal recursions that involve all three of the regimes of individuation and brings them together in new unities of operation.

NOTES

1. I will address these concerns in chapter 8.
2. It is interesting that Simondon chooses the pairing sociologist-psychologist in this context given that it is exactly this pairing he understands as the over-determined poles of the transindividual relation that he targets as requiring axiomization. That he has done so indicates the important place that technology plays for him in his proposed reconstitution of the human sciences.
3. This thesis of Wiener's is discussed in the first chapter and is readily found in Wiener's key works *Cybernetics or Control and Communication in the Animal and the Machine* (1965) and *The Human Use of Human Beings: Cybernetics and Society* (1989).
4. Simondon actually uses the example of the Guimbal turbine, which is used for the generation of electricity in dams. It offers a particularly clear example of concretization in that the dual functions of energy generation and heat dispersal are both carried out by the same flow of water through the turbine. The oil within the casing is also pluri-functional as due to its high pressure it prevents leaking as well as helps to grease the mechanism.

5. In his brief discussion of affective modalities in the essay 'Technical Mentality' (2009) Simondon identifies some differences between artisanal and industrial technology at the level of the affective engagement each typically involves.

6. For a particularly confused contemporary example of theorizing the aesthetic relation between technology, nature and culture, see the recent work *Technobiophilia: Nature & Cyberspace* (Thomas, 2013). Although initially claiming a definition of nature that is inclusive of all scientific and cultural production, Thomas also claims a clear distinction between the natural world and contemporary technical environments, the inhabiting of which calls for the experience of the natural world (now conceived more in line with that which is native) to soothe human spirits. This human spirit contains an essentialist 'innate tendency to focus on life and lifelike processes', which Thomas argues humans need to find in their technological use. That she also stipulates physical phenomenon such as streams of water as being vital (or at least lifelike) just adds to the confusion. All in all what we have here is a humanism stagnating in a romantic hippy ideal that reverts to type by providing a self-help manual to help the liberal self come to terms with technology-induced anxiety through a revised version of the escape to the 'natural' environment.

7. In Simondon's ontology we could describe this as the tension between the operation of the pre-individual ground with singular individuations.

Chapter Seven

Simondon, Latour and Stiegler

In this chapter I will contrast Simondon's ontogenetic approach for theorizing technology with the work of two important contemporary theorists: Bruno Latour and Bernard Stiegler. Latour has been chosen because his influential actor network theory shares several goals with Simondon in that it attempts to construct an ontogenetic, relational and realist account of technology. As well as identifying in what ways Simondon's work differs from their work I will begin to explore how it might offer another approach for thinking about contemporary media technology.

BRUNO LATOUR

A consequence of Kant's Copernican revolution is that organic products, unlike other entities, are experienced as self-organizing via the category of causality alone, and so don't conform 'to the constitution of our faculty of intuition' (CPR Bxvii). As a result of this inability to understand natural kinds by the conditioning *a priori* laws of intuition alone, non-determinant regulative principles are required. It is Kant's Copernican revolution that is understood by Bruno Latour as a defining moment in the history of modernism for it is with Kant that:

> Things-in-themselves become inaccessible while, symmetrically, the transcendental subject becomes infinitely remote from the world. (Latour, 1993: 56)

Unlike Kant's problem concerning the judgement of the natural purpose of organic products from the perspective of a transcendental subject, Latour's problem concerns the character of objects in general, given their dependence on just such a subject. He thereby undertakes a reorientation of Kant's antinomy from being one between the organic and mechanical to one between the natural and the artificial. Additionally, for him, objects are not differentiated according to this distinction but can be quasi-objects, which are objects that are partly social (artificial) and partly natural; rather than organic products, it is technological objects in particular that he is referring to here. To understand what the significance of these quasi-objects is we must explain that Latour's account of modernism describes two different sets of practices: purification and translation.

The practice of purification is that which 'creates two entirely distinct ontological zones' (Latour, 1993: 10), which are of non-human nature and human culture. The importance of Kant is that he is the first philosopher to perform the complete separation of these zones with his Copernican revolution. For Latour the dichotomy the work of purification creates describes the 'modern critical stance' (Latour, 1993: 11). This work has three approaches for explaining the world: naturalization, socialization and discourse. The first two approaches reside on opposing poles of the dichotomy with scientific naturalism, on the one hand, explaining the world in terms of mechanism and the laws of nature and, on the other, the human cultural and social explanations from whence freedom, intentionality and values are situated. The first dichotomy of modernism is defined by the incommensurability of these approaches.

Latour describes how this incommensurability is apparent in the problems social science faced when it tried to 'do for science what Durkheim had done for religion' (Latour, 1993: 54), which was to subsume it within sociological explanation. In attempting to do this it found itself face to face with the uncomfortable proposition that this would require the dissolution of faith in the existence of objective reality as described by the hard sciences. The impossibility of the subsumption of explanation to either side of the nature-human dichotomy thus left the modernizing project with a dualism in which explanation for any aspect of the world must be translated to one or the other of its sides.

The third approach undertaken by modernism is that of discourse, which instead of seeking explanations located at either of the poles of purification, focused on one of the mediators purification utilized, that is, language. In this approach the operation of discourse becomes the prime site for explanation through such practices as semiotics and deconstruction. Although situated between the nature-human divide it actually fails to describe the mediations of quasi-objects but instead charts the mediations of language itself, given the assumption of the withdrawal of nature, on the one hand, and the social constitution of linguistic practises on the other. Yet language as a natural capacity of rational animals marks a certain 'quasi-objectality' in it.

The second dichotomy describes the distinction between the practise of purification and the practises of translation. The latter describes the mediating processes by which quasi-objects are both constructed and influence one another. These quasi-objects are hybrids of nature and culture, which operate in relations of mediation called networks that describe the connections between the various things that come into relation to transform a given situation. For example, if giving an account of the damage to the ozone layer, one would 'link in one continuous chain the chemistry of the upper atmosphere, scientific and industrial strategies, the preoccupations of heads of state, the anxieties of ecologists' (Latour, 1993: 11). Such an account differs from that given by the work of purification as it mixes that which is supposedly incommensurable, namely, naturalized and sociological accounts.

The paradox of modernity is that as the number of these hybrid quasi-objects have increased through translation, the further apart have grown the poles of purification by which explanations of them are achieved. Modernism has attempted to ignore this paradox by denying the importance of these hybrids as realities in themselves. Its explanations of them aren't found in the mediations between them as real things but in the 'pure forms' (Latour, 1993: 78) of explanation developed via purification such as 'epistemes, mental structures, cultural categories, intersubjectivity, language' (Latour, 1993: 57). Instead of developing explanations by tracking mediations the purified account operates through these 'intermediaries', which sort phenomena to either the pole of nature or that of society.

What's more, rather than these two practices of purification and translation just occurring transversally to one another they are actually linked as the progressive work of purification enables the development of ever more kinds of quasi-objects in the form of technologies through the industrial application of purified natural science. Additionally these technologies, especially in the form of scientific instruments, enable the development of purification.

Despite this, Latour maintains that the modernist project cannot itself account for the 'irruption of objects into the human collective' (Latour, 1993: 21), for although the moderns obviously accept that there has been an increase in technological objects (translation) and a nature-social division (purification), they 'have never been explicit about the relation between the two sets of practices' (Latour, 1993: 51). It is the making clear of this relation that Latour terms non-modernism.

NON-MODERNISM

How then can modernism be overcome? How can we return technology to its 'ontological dignity' (Latour, 2002: 252)? How can we resolve the paradox of the modernist constitution?

Latour's response is to look for inspiration in the method of anthropological ethnography, which is capable of describing the entirety of nature-culture as simultaneously real, social and narrated rather than as reducing or 'purifying' the one into the other:

> The ethnologist will certainly not write three separate books: one dealing with knowledge, another with power, yet another with practices. She will write a single book. (Latour, 1993: 14)

As such Latour proposes that a method is required that, like the anthropological method, will take into account all the relations involved in defining any situation it studies. This involves stressing the relations between both the practices of purification and translation and in doing so 'accommodate the hybrids and give them a place, a name, a home, a philosophy, an ontology and . . . a new constitution' (Latour, 1993: 51).

The method Latour develops is to begin any investigation from the hybrids themselves and trace the network of their relations without the need to purify any explanation of what is discovered. He describes this proposal as a 'Copernican counter-revolution' (Latour, 1993: 79), which takes the quasi-objects and mediators as the starting point and focus of any investigation, and from which any notions of objective nature or subject/society emerge rather than being the transcendental conditions for understanding in the first place:

> If we seek to deploy the Middle Kingdom for itself, we are obliged to invert the general form of the explanations. . . . The explanations no longer proceed from pure forms toward phenomena, but from the centre towards the extremes. (Latour, 1993: 78)

Latour echoes Simondon in his proposal for a theory based on genetic development and the requirement for a new ontology, although his solution is considerably different. At the heart of Latour's ontology is the notion of the actor or actant. For him all quasi-objects and quasi-subjects are actors in the sense that they all have agency:

> I suddenly understood that the non-human characters had their own adventures that we could track, so long as we abandoned the illusion that they were ontologically different from the human characters. The only thing that counted was their agency, their power to act and the diverse figurations they were given. (Latour, 2012: 6–7)

One could perhaps summarize Latour's project, with apologies to Husserl, as 'to the actors themselves'. But unlike the phenomenologist Latour proposes a method of description that gets to the *reality* of the actors under consideration and not the reduced experience of them, which makes phenomenology merely a purified approach.[1]

By agency Latour does not mean that all technical objects are imbued with intentionality, rather that when actors come into relation with one another they transform one another. They have a 'relational materiality'. Thus Latour is able to define an actant with the broad definition of:

> Any element which bends space around itself, makes other elements dependent upon itself and translates their will into a language of its own. (Latour and Callon, 1981: 286)

It is through this process of translation or mediation that actors are both formed and give form to other actors. Thus if we are to describe an actor we must necessarily also describe the network of relations it has with other actors. Latour's method consists in tracking the network of actors through their mediations, for this is how reality is constituted and transformed. We must not mistake this reality for the purified descriptions that have previously been given of it. Due to the modernist nature-society split such descriptions are necessarily always partial, due to falling either side of this split, or reduced just to the mediation of discourse. Latour describes the poles of nature and society as huge tectonic plates that have emerged from the magma of mediating activity below (Latour, 1993: 87). They should not be seen as causes but as effects—it is this reversal that describes his Copernican counter-revolution. For if we follow his method we will find that the actors are not purified entities but that different aspects of their description cross the nature-society divide.

To be true to reality, just like the anthropologist, we must give as full a description as possible using as many practices as possible even if that means our descriptions are simultaneously natural, social and discursive. Such practices are just ways of carving up the world into purified chunks, whereas what Latour is aiming for is a more complete account by stressing the relations between them:

> We have both [moderns and non-moderns] always built communities of natures and societies. There is only one, symmetrical anthropology. (Latour, 1993: 103)

There are some similarities between Latour's and Simondon's projects. Both have the avowed aim of demonstrating the importance of technology in response to it being either misunderstood or ignored due to reasons that are ultimately metaphysical. Simondon's *bête noire* is hylemorphism (among others, of course), while for Latour it is the purification performed by modernism. They both, also, aim at a genetic tracing, an account of the individuation, of the social, and the technological. But despite both proposing an understanding based on relations their differing ontologies lead to dissimilar consequences.

AGENCY AND CAUSALITY

As I have already mentioned, it is the actor that is central to Latour's proposed ontology and what makes an actor an actor is its 'power to act' (Latour, 2012: 7). In this section I will interrogate what grounds Latour's ontology of actors and if it is able to offer a credible realism. As already noted, Latour's concept of translation is both relational and causal in nature. Actors are such that they are permanently involved in processes of mediation with other actors. In fact Latour develops a range of terms he uses to describe the various processes of mediation in which actors are involved: for example, delegation, prescription, fold, inscription, translation.

Thus in describing a network using actor network theory (ANT), a theorist might utilize a number of different terms for the mediations involved. For example, a washing machine translates or inscribes the actions of hand-washing clothes into a machine process. In turn the machine, perhaps due to its speed and ease of use, might well prescribe back to its users new behaviours related to washing clothes or new values regarding hygiene, cleanliness or housework. What would be attempted is a description of how such mediations involve causality involving different types of actors (for example, people, clothes, machines, detergent, values). As such the explanation is not just cultural, physical or discursive. It involves aspects of all of these. In this network of different kinds of mediation Latour is pointing to something important. It is true that different kinds of things can be involved in related causal processes. However, he fails to offer a convincing account of the causal processes involved. The various words used elide this crucial part of the story. Latour is right to talk of hybrids and mixing but his ontology remains too underdetermined to offer a convincing account of how this actually operates. As such we would liken this to the 'obscure zone' Simondon identifies in those principles of individuation, such as hylemorphism, which presuppose the operation of individuation having already taken place by specifying the resultant terms as part of the explanation.

One reason for this is that Latour's anthropological method actually avoids developing a truly ontological approach (such as Simondon's) and instead just draws together explanations garnered from a range of practices in the hope that they will be sufficient for a consistent explanation. The problem with this is twofold. First, this does not give us a causal story at a granular level of detail. Second, it is still not clear how the different explanations may be connected without a coherent ontology. In *We Have Never Been Modern* it seems that it is mediation that is to offer this explanation: 'The world of meaning and the world of being are one and the same world, that of translation, substitution, delegation, passing' (Latour, 1993: 129).

However, this description seems to sidestep the reality of mediation in that it claims to bring meaning and being together without detailing how. In short, Latour's network anthropology lacks a theory of ontogenesis that describes not just how actors emerge in the first place but also how they relate causally. At its heart Latour's anthropological approach retains the modernist division it rails against because it fails to link the different practices of explanation into a unifying axiomatic of ontogenesis. It is precisely this that makes Simondon's account so much more comprehensive.

In his later work Latour appears to be attempting to address this lack by developing an account of multiple modes of existences, which he calls a multirealism:

> My hypothesis is that each of these modes makes it possible to respect, in the empirical areas I have pursued up to now, a certain tonality in the experience, the felicity or infelicity conditions particular to each case, and especially (here is where things become dangerous) a specific ontology. In fact, each mode requires us to encounter distinct beings which must be addressed in their own languages. The classic question of philosophy, 'what is the essence of technology, science, religion, and so on?' then becomes 'what are the beings appropriate to technology, science, religion, and how have the Moderns tried to approach them?' (Latour, 2012: 1–2)

This approach sees the continuation of Latour's anthropological method. His hope is to return to experience in order to try and identify these different modes and the theories appropriate to them. As such it sounds like a form of multidisciplinarity situated within the unifying rubric of the descriptive science of anthropology.

A question worth asking here is: Does such an approach live up to the claims of his Copernican counter-revolution? For it would seem to be the various empirical areas of explanation that are determining of ontology rather than such explanations arising from the hybrids themselves. This would seem then to be a kind of reformed Aristotelian substance, independent of its predicates, developed as a range of modes with discourses appropriate for each one. However, as is often common with cross-disciplinary exercises, there is little that connects them *as such*: the genesis and actuality of objects themselves is still lacking.

Such an exercise appears as a new system of purification, sorting entities into these different modes through intermediation (that is, discursive practice) rather than the proposed tracking of networks of mediators. As such this runs contrary to the kind of counter-revolution we understand Simondon as developing: that of tracking the genetic development of individuation from the pre-individual onwards in such a way that the connections between epistemology and ontology remain coherent and robust.

With such attention given to discourse it is perhaps unsurprising that Latour has sometimes been mistaken for a social constructionist, something that frustrates some of his followers. For example, Cordella and Shaikh bemoan the case that too many researchers use ANT for an interpretivist approach when they should 'allow the actants to "speak" for themselves, and not put words in their mouths' (2006: 17). To do this, they add, researchers should use 'the ontological dimension of ANT' (2006: 17), although it is not clear how distinct this ontological dimension is from the discursive, owing precisely to Latour's anthropological method. Additionally, Latour also completely omits the question of ground and individuation, as for him the network of actors always already exists.

AGENCY AND THE HUMAN

Latour's attribution of agency to all entities raises further issues. First, he fails to give an account of agency that distinguishes this property between the various actors capable of demonstrating it. As noted above agency is that power an entity has to act, but what kind of power this is for any given actor is concealed behind the generalization of terms such as mediation and translation. It is instructive to examine Latour's brief discussion of the human in *We Have Never Been Modern* to expand on this concept. Like Simondon, and as a constructivist, it should not be surprising that Latour doesn't believe in a human essence. Instead it is clear that like all other actors the human is a mediator. That is to say that it shouldn't be defined by freedom, at least in the Cartesian sense, as this is to return to the subject-object dualism that beset modernism. Instead humans are entangled with all other actors in a process of shaping and being shaped:

> The human is in the delegation itself, in the pass, in the sending, in the continuous exchange of forms. (Latour, 1993: 138)

If we are generous we might be able to forgive the implicit hylemorphism in this statement. Perhaps Latour could be interpreted as meaning 'exchange of forms' in the sense Simondon uses terms such as modulation of information. However, due to the lack of detail Latour provides on this issue, we can't be sure. Neither does his definition of the human help clarify the situation: 'A weaver of morphisms—isn't that enough of a definition?' (Latour, 1993: 137).

If we are to follow Simondon here we must reply in the negative. Man is not just one mediator among others. To claim this is to do a disservice to both human and non-human. Simondon's project, especially in *The Mode of Existence of Technical Objects*, is similar to Latour's in that both seek a more developed recognition of the technical. However, what Simondon develops goes deeper than just recognizing some level of technological agency in an expanded anthropology.

Simondon views anthropology as being problematic, and in some respects his and Latour's concern with this domain share common concerns. Rather than being the solution for a dualistic modernism, however, Simondon understands anthropology in terms similar to that by which Latour understands modernism itself. For Simondon anthropology denotes both those attempts to define the human as either essentially psychic (for example, psychology) or social (for example, sociology) and in doing so as distinct from other animals, as well as that tendency to reduce technology to just a means for humans ends. As such it broadly describes the two dichotomies that Latour describes as modernism, that is, those between nature/culture and nature-culture/technology. I say broadly as obviously Simondon's ontology accounts for the difference between and actuality of nature and culture differently from Latour.

Can we draw from this a criticism of Latour's use of anthropology to resist a modernism that contains those two dichotomies that also constitute anthropology for Simondon? As I hope is becoming clear Simondon's broad ontological investigation into individuation takes him far beyond a reinvigorated anthropology (in Latour's sense).

Simondon's project involves what Barthélémy (2010) calls a 'difficult humanism', which has as its goal a response to the two tendencies described above. On the first point it should be clear from previous chapters that Simondon doesn't recognize a distinction between humans and animals on the grounds of psychology, for he fully allows that animals can find themselves in psychical situations because psychic individuation is a development of the vital regime and thus necessarily not restricted exclusively to humans. On the second point, as we have also described, Simondon's understanding of technics goes far beyond seeing it merely as a means because it has its own 'mode of being' and remains always in excess of its particular uses.

What distinguishes Simondon even more from Latour, however, is the positive aspect of the former's project, which calls for an understanding of the relations that operate between culture, nature and technology. It is charting this in a genetic manner that constitutes his new encyclopaedic project, which has as its goal the overcoming of the alienation described in chapter 5.

For Simondon humanism is something that requires frequent reinvention, for the human is constantly becoming as it finds itself in new situations, not least in relation to technology. With the development of both thermodynamic and informational technologies Simondon discerns the need for a new thinking of the culture-technology relation so as to overcome alienation. This also requires an understanding of the part technology plays in establishing trans-individual relations as well as the development of the role of conceptual ideation through analogy in invention. As Barthélémy succinctly describes it,

> Simondon reminds us that the whole of Encyclopedism aimed to liberate man from a determinate alienation, and the question today is one of combating a new form of alienation (new because it is machine-induced) with a new Encyclopedism (new because it is genetic). (Barthélémy, 2010: 245)

Due to the flattening nature of the reduction of all entities to actants, as well as a lack of clarity regarding causality, Latour's philosophy of technology is remarkably unambitious. From his perspective there isn't really too much to be worried about. The modernist fears regarding technologies are merely the result of purified myths. For example, he swiftly dismisses Ellul's concerns:

> Protecting human beings from the domination of machines and technocrats is a laudable enterprise, but if the machines are full of human beings who find their salvation there, such a protection is merely absurd. (Latour, 1993: 124)

Once it is understood that there are just networks of actors then such a substantive understanding of technology must be refused. In truth there have been no epochal jumps just the gradual prolongation of networks and 'the innovation of lengthened networks is important, but it is hardly a reason to make such a great fuss' (Latour, 1993: 124).

For Latour the difference between the non-moderns and the moderns is down to the length of the networks the latter have constructed. That these networks are sometimes thought of as totalities is a mistake: 'Since this enlistment of new beings had enormous scaling effects by causing relations to vary from local to global, but we continue to think about them in terms of the old opposite categories of universal and contingent, we tend to transform the lengthened networks of Westerners into systematic and global totalities' (Latour, 1993: 117).

There is a tendency in Latour's work to resist systematization and reduce everything to actants in networks. This results in a neglect of the operative nature of reality that Simondon describes in his allagmatics. An important consequence of this is that it leads to an uncomplicated and unquestioning ethical perspective on technology in which what is most valued is the continuing development of fresh mediations:

> Every concept, every institution, every practice that interferes with the continuous deployment of collectives and their experimentation with hybrids will be deemed dangerous, harmful, and—we may as well say it—immoral. The work of mediation becomes the very centre of the double power, natural and social. (Latour, 1993: 139)

Unlike Simondon's more balanced aesthetic approach, which enables an identification of man-technology and technology-nature combinations that are alienating or catastrophic, Latour seems reluctant to pass judgement as such concepts are too universal. Besides, as a mediator, technology is too disruptive and contingent to be able to be the subject of morality. Technology always presents us with unforeseen consequences that are impossible to judge prior to the event and thus it should not be subordinated to morality: 'As it is often said, morality is less preoccupied with values than with preventing too ready an access to ends' (Latour, 2002: 257).

Instead Latour is content to settle for a democratic solution to questions of which networks to build, but the *demos* he calls for is of a new kind in which 'things' also have a voice, where *all* actants are represented. This begs the question of *how* things are represented, the answer to which would seem to be based on Latour's now-broadened anthropology. But isn't this problematic? Can the things really be said to represent themselves when the method by which they do so remains anthropocentric, however well intentioned? At least Simondon is honest in that he explicitly seeks an aesthetic balance between culture and technology and technology and nature at all levels. Latour's democracy, like all democracies, is open to abuse by those stronger actors who wield more power in 'tests of strength'.

Latour's democracy and networks, in drawing attention to the reality of technology, simultaneously feel like its defanging. Although, just like Simondon, he recognizes that technology is in some sense 'made of subjects' (Latour, 1993: 196), he understands this as its domestication whereas Simondon understands it as a problem in need of resolution (much like Stiegler sees technology as *pharmakon*, as we will now see).

BERNARD STIEGLER: *TECHNICS AND TIME*

Given the breadth of Bernard Stiegler's oeuvre it is difficult to offer a summation of his overall project. In this section I will aim to highlight those aspects of his project that I understand as either directly influenced by Simondon or that seek to extend Simondon's project to include contemporary digital technology. In Anglophone culture Stiegler is best known for his *Technics and Time* series of books. In this series he develops most of the core concepts that continue to inform his more recent politically engaged work.

As with both Simondon and Latour, one of Stiegler's central concerns is the marginalization of technics in philosophical thought. In the first book of the series Stiegler seeks to counter this tendency by demonstrating the importance of technics for the constitution of humanity, in particular that this involves the fundamentality of technics for human temporality. Stiegler argues that the human lacks both essence and origin. It is technics, understood as a supplement in relation to this lack, that drives the process of the co-invention of the technical with the human:

> The technical inventing the human, the human inventing the technical. Technics as inventive as well as invented. This hypothesis destroys the traditional thought of technics, from Plato to Heidegger and beyond. (Stiegler, 1998: 137)

Simondon's work is crucial for Stiegler's description of the genealogical development of this co-constitutive relation because it enables an account of technics as quasi-independent, due to it having its own mode of individuation, while also being co-related with the human. This co-originary relation means that technics plays a prominent role in all aspects of human affairs. It will be one of Stiegler's aims to give an account of the multifarious ways that the human and technics interact in what he calls a 'general organology'.

As Stiegler mentions in the quotation above, this relation also challenges traditional ways that technology has been thought. It does so by undermining those critiques of technology that understand technology as destroying a more originary mode of relation of the human with the world. Thus Heidegger's claims that technology is a mode of truth that endangers man from experiencing the call 'to a more primal truth' (Heidegger, 1977: 28) is undermined because, for Stiegler, there is no more fundamental relation with the world that doesn't involve technics.

The co-constitution of the human with technics is also a process of exteriorization: human subjectivity is constructed in a process of exteriorization in relation technics.

> There is no interiority that precedes exteriorization, but to the contrary exteriorization constitutes the interior as such, that is to say, distinguishes and configures it in the very course of what Leroi-Gourhan describes as a process of exteriorization where this configuring distinction, which is constantly displacing itself, each time setting up new relations between the psychical individuals and the collective ones—new processes of the formation of psychical and social individuation, in the sense Gilbert Simondon confers to this expression while stipulating that memory is the 'associated milieu' of this individuation. (Stiegler, 2010a: 70)

Simondon's influence is clear here with the importance Stiegler gives to the notion of an 'associated milieu', however, we can also see how the exteriorization process is also reminiscent of the process of Simondon's image-cycle. A key difference between the two thinkers is that Simondon doesn't describe technics as being originary for humanity; for him the human inhabited a non-technical magical world prior to technical development. This also means that technics doesn't constitute the pre-individual for Simondon, as it does for Stiegler, for whom the history of the human is also the history of technology and vice versa. As such he cannot describe a *real* history of technology that is in any sense independent of, even if co-extensive with, recorded human culture. His is therefore necessarily an anthropological theory of technology in contrast to Simondon's more precisely described technological and ontogenetic account.

Since its establishment with the development of flint tools two million years ago, the history of the exteriorization process has simultaneously been the history of the individuation of the mental *(the 'who')* in relation to that which it is exterior *(the 'what')*, which for Stiegler is technics. To understand the importance of exteriorization for the development of the mind for Stiegler one also has to understand his concept of epiphylogenesis.

Developed from both anthropological and zoological sources, epiphylogenesis is the theory that human development is not just one of genetics but is deeply implicated with technics as 'the pursuit of life by other means' (Stiegler, 2009: 163). That is, that human development is profoundly influenced by the environmental presence of technics, of the sedimentations of epigenesis in its milieu, which leads to a 'break with pure life, in that in the latter, epigenesis is precisely what is not conserved' (Stiegler, 1998: 140). The preservation of the epigenetic, which also has its own developmental dynamic (the technical tendency), enables new forms of reflexivity and anticipation that help overcome the retentional finitude of the biological organism as well as enable it to develop new temporal relations.

By anticipation is meant the 'realization of a possibility that is not determined by a biological program' (Stiegler, 1998: 151). Stiegler also uses the term *protention* to describe this anticipatory capability that technics helps to establish through being an intergenerational mnesic support.

Thus the history of technics is also the history of the development of human memory via exteriorization, an exteriorization that Stiegler, following Husserl's classification, calls tertiary retention. Stiegler develops the concept of tertiary retention in *Technics and Time, 3* by developing Husserl's notions of retention and protention from *Logical Investigations.* In his analysis of the temporality of consciousness Husserl identifies a temporal object as that which 'is constituted only in its duration' (Stiegler, 2011: 13). The example used is that of a melody, which is constituted through time and as such 'manifests itself in disappearing' (Stiegler, 2011: 14). Such temporal objects maintain phenomenological unity over the course of their duration and Husserl names the maintenance of this unity primary retention. Without this retention the melody would not be experienced qua melody but as a series of instants. Primary retention is the phenomenological maintenance of the past part of the melody for its listener during the duration of its performance.

Secondary retention refers to the effect primary retention has on subsequent experience. That is to say that the retention of the previous experience of a temporal object has a conditioning effect on ensuing experience. This effect is recognizable in that successive experiencing of the same melody are never experienced as identical. The previous hearing of the melody, as well as any other music, will have changed present experience.

To these two retentions Stiegler adds tertiary retention that is 'an artificial memory presented in a support medium' (Stiegler, 2011: 23). This refers to the already-there of material culture (including technical objects) that acts as an 'intergenerational support of memory' (Stiegler, 2010a: 9).

Stiegler argues that primary retention is always already dependent on secondary retention, which in turn relies on the already-there of tertiary retentions, which is technicity. Therefore there is never a separation of the *who?* from the *what?*, the living from its non-living milieu.

It must be understood that epiphylogenesis doesn't just mean a prosthetic extension of memory.[2] The profundity of epiphylogenesis lies in it being a transductive process in which the *who?* and the *what?* co-individuate. This means that prior to the process of exteriorization there is no interiority:

> the issue is therefore neither that of an interiority nor that of exteriority—but that of an originary complex in which the two terms, far from being opposed, compose with one another (and by the same token are posed, in a single stroke, in a single movement). (Steigler, 1998: 152)

Therefore what are commonly described as mental phenomena, such as gesture, language, numbers and memory, are the result of an exteriorization process. Stiegler recognizes the importance of the body and gesture for the exteriorization process as well as for the phenomenological constitution of time. A gesture is anticipatory in that it only exists through the possibilities created by that which is external (the tools and prosthetics that constitute tertiary retentions) and which increases the possibilities for action in its milieu.

> There is no anticipation, no time outside of this passage outside, of this putting-outside-of-self and of this alienation of the human and its memory that 'exteriorization' is. (Steigler, 1998: 152)

Similarly the development of and capacity for mathematical manipulation also arose from a process of external manipulation that has become mirrored 'internally':

> In point of fact, number in general can only be conceived of as being determined within a system of traces, any notation system constituting itself through the external manipulation of symbols: there is no mental calculation not resulting from the secondary interiorization of a calculation by symbolic manipulation, that is to say through manual behaviours. (Stiegler, 2011: 52)

In Stiegler's description we have a detailed account of the development and extent of the co-individuation of the *who?* with the *what?*, which demonstrates that the mental always stands in a constitutional relation to an exterior.

In particular the phenomenological richness that we experience, the complexity of language and numeracy, is explained by the anticipatory relationship maintained with an exterior milieu composed of technics.

In this description of exteriorization we discern a very similar account to that given by Simondon of invention in his theory of the image-cycle. Although Stiegler, like Latour, relies more on anthropology (Leroi-Gourhan) and, unlike Latour, on phenomenology (Husserl) than the biological sciences, which Simondon utilizes, there is a commonality in their projects in that both wish to demonstrate the causal relation between ideation, biogenesis and the physical world. Additionally, as we will see when considering his political thought, Stiegler also utilizes Simondon's notion of transindividuation in order to explicate the social implications of this co-constitutional relation.

Stiegler pays particular attention to developing an account of the operation of media technologies in this co-constitutional relation with a focus on the development of time consciousness in relation to orthography, cinematography and digital networks. From the perspective of thinking Stiegler as a post-Simondonian thinker[3] we can understand him as describing a new kind of alienation in relation to these contemporary technologies that builds on Simondon's own account of alienation.

FOR A NEW CRITIQUE OF POLITICAL ECONOMY

In *Technics and Time, 2: Disorientation* Stiegler develops his account of the industrialization of memory due to the convergence of modes of cultural production and consumption utilizing digital networks with capitalist economic imperatives. He sees in this synthesis the danger of a 'psychopower' that subordinates social and cultural individuation to determination by speculative economic concerns. To understand the developmental logic of this over-determination I will briefly present the argument of probably his most polemical work to date, *For a New Critique of Political Economy* (2010a).

A concept crucial for understanding the industrialization of memory is that of grammatization, which is 'the process through which the flows and continuities which weave our existences are *discretized*' (Stiegler, 2010a: 31). Grammatization involves the technical exteriorization of the various kinds of memory. Thus the reproduction of gestures by automated machines, such as, for example, with powered looms and punch cards, is one example of grammatization in which the worker's gestures are discretized and recorded in the holes of punch cards. Stiegler describes this kind of technical object as mnemotechnics.

Technical history is, for Stiegler, a history of the grammatization of the different kinds of memory by various types of technology. Thus both language and numeracy are types of grammatization that support different kinds of psychosocial individuation. Stiegler also follows Simondon in his understanding of proletarianization as that which occurs when the worker's knowledge is subsumed into the technical object:

> The proletarian, we read in Gilbert Simondon, is a *disindividuated* worker, a laborer whose knowledge has passed into the machine in such a way that it is no longer the worker who is individuated through bearing tools and putting them into practice. (Steigler, 2010a: 37)

This loss of knowledge is described by Stiegler as a loss of *savoir-faire.* It is the synthesis of converged digitized media with a neoliberal mode of economic production that Stiegler sees as posing an altogether more widespread form of proletarianization, which extends from alienating the knowledge of the worker to all other forms of knowledge including the cultural.

For Stiegler the progression of the grammatization process into the social and cultural entails an even more profound loss of knowledge, that is, the proletarianization of *savoir-vivre.*[4] Stiegler's argument here reiterates aspects of various critical engagements with contemporary digital culture including those of immaterial labour (Lazzarato, Terranova), attention economy and neuroscience (Carr) and the control society (Deleuze).

Stiegler's fear is that just as the grammatization of workers' knowledge and skills led to work becoming disindividuating labour and with workers reduced to being mere servants of technical individuals, so the grammatization of culture in the service of marketing will lead to a hyperindustrial consumerist control society.

> Within the current capitalism typical of control societies, the function of culture has been reduced to socializing production by standardizing consumer behaviour, culture thereby becoming the agent par excellence of this control. (Stiegler, 2011: 26)

The over-determination of contemporary Western culture by corporate marketing in the service of capitalist economics is therefore leading to social homogenization bereft of any transindividuating capability.

The transindividuating potential of the social is undermined by a short circuiting of the collective desire for long-term shared aims into individualized and easily satisfied short-term drives. Given that all individuals undergo the same conditioning, the inevitable result is the production of a monoculture of neoliberal subjectivity.

Where Simondon sought to address the problem of culture lagging behind the development of technology in such a way that it misunderstands it and sees it as a threat, Stiegler points to the advancement and domination of economics as the main problem:

> In the twentieth century, however, the economic system having taken a step beyond all the other systems, and being charged with the task of unifying them *by finitizing them,* that is submitting them to a process of generalised 'monetization' . . . it is *infinitive consistence* (the law of desire), constituting the condition of any genuine co-individuation of the three organological levels, which finds itself destroyed. (Stiegler, 2010a: 106)

Such a situation is necessarily self-destructive. If all desire is attenuated then it becomes short-term, which means the destruction of desire regarding a long-term horizon. Such short-termism ultimately leads to a '*liquidation of social relations*' (Stiegler, 2010a: 57). Here Stiegler's critique shares aspects of Marcuse's critique of capitalism due to desublimation,[5] in that desire no longer aims for the infinite (though it may never be attained), but as drive is instead easily, if unsatisfactorily, sated through consumption.

The question of grammatization thus leads to the need for a new political economy that must address the problems of the industrialization of protention and subsequent dissolution of libidinal energy. In order to do this Stiegler proposes the need for a general organology, which bears a considerable resemblance to Simondon's allagmatics in that it is a theory of 'the transductive relations between the three levels' (Stiegler, 2010a: 117), which are the psychosomatic, the technical (or pharmacological) and the social.

Although, as the name suggests, Stiegler generally proposes his organology as describing the relations between various types of organs (bodily, artificial and social), he is clear that there 'are tendencies and counter tendencies proper to each of the three organological levels' that come into transductive relation. As such, like Simondon, Stiegler's investigation aims for a processual account of and between psychosomatic, social and technical structures.

Stiegler is also clear that although these levels have distinct 'tendencies', these are causally interwoven. Therefore the technological level can be understood as *pharmakological* because its influence on the psychosomatic level can lead to either psychic proletarianization and thus social disindividuation or psychic individuation leading to social transindividuation. The question of a political economy involves then the tracing of the relations between organological levels in order to understand the pharmacological dimensions of any mnemotechnics and the possibility for any subsequent therapeutics. By this Stiegler refers to the ways by which the poisonous aspects of any *pharmakon* can be remedied and psychic and transindividual individuation encouraged. He identifies a number of areas where he sees the digital networked mnemotechnics of the contemporary technical level as having a toxic effect on both the psychosomatic and social levels. For example, a core area of concern is with the purported transformation in the nature of attention, which is particularly experienced by the younger generations due to their excessive contact with these technologies. This has led to a reduction in the ability for 'deep attention' and with this the ability for the maturation of thought and socialization. This problem is exacerbated by the short circuiting of desire, which the penetration of economic markets into these networks causes. Such a combination has led to the increase in attention disorders such as ADHD.[6]

Stiegler sees in such transformations a need for a 'battle for intelligence', which he proposes should be undertaken through a 'psychopolitics' that, since the psychic is not reducible to the life of the concept, implies a 'noopolitics' that, following Simondon's observation that a psychic individuation is simultaneously also transindividual, politicizes the location of the concept.

Unlike Latour's democratic 'parliament of things', Stiegler's political solution to what he sees as a catastrophic situation for humanity requires a range of strategies that correspond to different relations across organology. For example, one strategy is to circumvent those aspects of network culture controlled and exploited by capitalist interests. Stiegler describes this strategy as 'taking care' in that what is being fought is a '*struggle against the careless tendency* inherent in that *pharmakon* that is capital, and *thus to take care of the world*' (Stiegler, 2010a: 108).

The overall goal of these strategies is to establish a long-term common desire for a renewed techno-social project that aims to constitute what he calls 'long circuits' of transindividuation. Such long circuits are what he has understood as being short circuited by the current dominant forms of grammatization. What Stiegler has in mind with this is how a change in the dominant mode of recording, storing and replaying the various kinds of human memory (for example, gesture, language, photography, video) can disrupt and change social individuation in such a way that it acquires a new political and semantic character. This political project must ultimately be based in aesthetics, for the sensational basis for noetic life is indissolubly social and the foundation for the development of culture. In particular, what culture means is the establishment of consistences, that is, non-existent yet *consisting* infinitive ideas at which desire can aim: for example, the idea of justice. Such consistences are required to play a regulative role in sustaining a belief for the long-term stability of social individuation.

> There is practice and culture because there is ancestry and inherited obligations that, far from being the opposite of the freedom of singular time, are, as pre-individual funds, the condition of such freedom. This is what forms itself as—and forms—consistences. These 'forms', which are however wholly informed by the material constraints of tertiary retention permitting their stabilization and transmission, metastabilize themselves as a process of psychic and collective individuation. (Stiegler, 2011: 118)

Such consistences must neither be understood as theologically given nor reducible to calculation (and therefore to economy). The problem then is one of the maintenance of such consistences through changes in the technical aspects of organology when desire necessarily is originarily technical and any change in grammatization conditions in turn causes adaptations to psychosocial individuation.

There is thus a tension in Stiegler's thought between the required maintenance of consistences and the plasticity of composition that the history of epiphylogenesis describes. As such he reflects the Simondonian tension between invention and regulation. Stiegler's requirement for consistences can also be understood as a plea for the conservation of enlightenment rationality and values. Certainly some of Stiegler's concerns about declining educational standards can be interpreted in this way. Like Simondon, Stiegler proposes an ontology based on composition and invention that understands as a requirement for transindividuation the need for these hyperstitional fictional consistences upon which collective desire can take aim:

> *Noetic* life is *intrinsically* fictive, fictional, and, as such, *to be decided*, decided *in the political economy* of the *libidinal and spiritual* economy that a city constitutes—it is deciding to *realize a fiction*. It is wanting to believe in a fiction: law, insofar as it is a difference we must *make*. Or to put it another way: it is to have *imagination*—or, yet again, *to invent*. (Stiegler, 2011: 147)

Once more what we find at stake is a matter concerning the nature of causality in that such questions ultimately concern the ability to predict and anticipate the nature of any invention. With Latour we witness a prohibition on prohibition in that the law is not to deny actors their ability to act. Stiegler is wary of such a *laissez-faire* attitude. For him the claim that the unpredictability of the technical tendency means 'that it is not possible to predict the technical future, nor is it possible to build any kind of political will or bring it into reality' (Stiegler, 2010a: 124) is based on a confusion. Although the technical tendency (its mode of individuation) may be impossible to predict, the *technical fact* can be negotiated as it requires 'compromises between technical tendencies and social systems, which are themselves organizations resulting from tendencies and counter-tendencies constituting them as metastable systems' (Stiegler, 2010a: 125). As such he is entirely consistent that noetic individuation is absolutely central for our ability to enter into such negotiations and we need to take care of our ability to think and communicate complex ideas as well as resist a decline into a drive-based libidinal economy.

So to what extent can Stiegler be understood as developing Simondon's project and in what ways does he diverge from it? Stiegler's project is undoubtedly a significant intervention in the philosophy of technology, which borrows some key concepts from Simondon (for example, metastability, transindividuation and aspects of the image-cycle). However, Stiegler uses these concepts for a different end from Simondon. Although he also proposes a kind of technological alienation, he does so via the reduction of the human-technology relation to one based on a phenomenological account of memory. Additionally, through his theory of the image-cycle Simondon does propose schemas that are understood as having a regulatory purpose; he does not propose these as fictions that must be *decided* upon as Stiegler does. Simondon's schemas are not invented through a process of the voluntary selection of fictions but through an ontogenetic process of epistemological and normative development. The difference is perhaps subtle, but where Simondon demands an immanentism of the individuation of ethics along with that of technics, Stiegler ultimately retains a transcendence of the enlightenment subject as the cultural guide of technical development. There is perhaps a negotiation to be had here regarding what aspects of the enlightenment inheritance are being proposed as needing to be taken forward and to what extent we understand Simondon's modernism as doing this.

As we have seen, Stiegler's organology shares some aspects of Simondon's mechanology but ultimately betrays Simondon's core insight that the technological mode of individuation is in a very real way independent and not a co-constitution of a *who?* with a *what?* Additionally Simondon does not hold that man and technics are born as one. How could this be if the primal magical mode of being is prior to the birth of technics? As such, unlike Stiegler, Simondon resists a philosophical anthropology.

Stiegler makes an important contribution for thinking about the psychic and collective implications of networked information technologies. Although he is correct to make a case for a new form of alienation or proletarianization, it can often seem in his writing that the toxic aspect of this technology is exaggerated at the expense of a more balanced approach. Often Stiegler seems inclined to accept the doom-laden prognostications of digital naysayers, such as Nicholas Carr, as universally applicable without allowing consideration for a more balanced investigation that a genetic approach should give. Although Stiegler presents a convincing narrative of the implosive decadence of contemporary society, his concentrated focus on just the human retentional economy rather than technologies relations with a broader conception of the human may be misleading. However, Stiegler's strength is also in that he sees far more clearly than Simondon the cultural and economic aspects at play in the conditioning of the techno-social. This is undoubtedly due to the social nature of much contemporary technology use. This does not mean that Simondon is wrong just that contemporary technology requires reconsideration.

NOTES

1. For Latour phenomenology is a furthering of the paradox of Kant's Copernican revolution in that it 'transforms a distinction, a separation, a contradiction, into an insurmountable tension between object and subject' (Latour, 1993: 58).

2. For example, as the *active externalism* of Andy Clark and David Chalmers's *Extended Mind Thesis* runs the risk of reduction to. In this thesis objects in the environment are seen as being functional extensions of the mind rather than strictly co-constitutional as Stiegler argues.

3. I'm aware of the limited nature of this perspective. It would be just as easy to think of Stiegler as post-Derridean or post-Heideggerian. However, for the purposes of this text, such a perspective seems justified.

4. By *savoir-vivre* Stiegler is referring to behaviour that is more cultural and general in kind and not that associated just with working practices *(savoir-faire)*. An example would be that of the development of the idea of 'lifestyles' as marketing creations to drive consumption.

5. Marcuse develops his ideas in several places, most notably in *One-Dimensional Man* (1991) but also in his essay 'The Containment of Social Change in Industrial Society' (2001) where he writes: 'repressive desublimation, is characterized by the contraction rather than the extension of erotic energy by its contraction to sexuality—that is to say by a contraction and reduction rather than strengthening of the life instincts' (Marcuse, 2001: 91).

6. Stiegler describes this connection at some length in *Taking Care of Youth and the Generations* (2010b). For example, 'The psychosocial state of the world is equally ubiquitously . . . being overtaken by a colossal deficit of attention, an immense neglect in the *global attention deficit disorder*, stemming directly from the proliferation of psychotechnologies that no political power can control' (57).

Chapter Eight

Towards a Theory of Media

Although Simondon was working prior to the establishment of the Internet as a significant media technology and although he didn't develop his work into regarding what, if anything, made media a suitable domain for investigation (beyond some brief writing on cinema), in this chapter I want to make the case for a Simondonian influenced media theory.

More specifically I want to argue that Simondon's overall project cuts across many of the areas of debate that concern contemporary media theory as it attempts to come to terms with networked media technology and thus helps create a unified perspective.

This should come as no surprise given that so much contemporary thinking on media also shares a direct lineage to cybernetics. As such we can contrast Simondon's reformulation of this project with the interpretations we find in such varied traditions as British cultural studies (one of the central founding statements being Stuart Hall's reworking of Shannon-Weaver), American post-humanism (with its cyborg heritage) and German media theory (Kittler relies on the non-semantic nature of the Shannon-Weaver model).

Through making these contrasts we will be able to tease out the position Simondon's work offers us in negotiating some of the abiding issues that have plagued media theory in ascertaining the relationships between nature, culture and technology. Central to unpicking this traditionally Gordian tangle will be Simondon's informational ontology, his allagmatics. Through this we shall be able to steer a course between technological determinism and constructionism but also address a series of other problems: What is the place of humanism? Have the so-called new media ushered in a new form of alienation? What new kind of unities or structures and their modes of operation require consideration?

What makes Simondon unique are his realism and his encyclopaedism, perhaps best understood as his goal of axiomatizing the human sciences. This gives his theory an expansive scope, free of anthropologism, although not free of complexity.

A central aspect of my proposal, which is implicit in Simondon, is to understand media as environmental and not just a means of sending messages between individuals. It is also about how this environment, of which we are a part, individuates and is regulated and that it requires an approach that is not merely cultural but also natural, technological and psychosocial. In short, it brings us to the complexity of operations occurring at numerous levels, of hybrid structures and operations shifting and morphing.

Where we saw Latour's notion of mediation as translation fail to give a truly causal account, instead combining a number of approaches in a broadly anthropological method, thus evading the stipulation of a coherent ontology, with Simondon I believe we have a notion of mediation that involves a realist account of causality involving technical objects as *mediators* between humans and nature.

The claim for understanding contemporary media technology as environmental relies on a number of properties that suggest a shift has occurred that prevents them from being reduced to just being understood through the technical schema of the network, but the requirement for also evincing what we see as the imbrication of levels of operations and operational unities in which we are embroiled. What is being indicated is the technization of the social milieu by technologies that don't merely transport messages but that modulate our very experience of the world, often eluding human intention altogether. Contemporary media technologies enter into their own operations, as indicated by the array of adjectives they require, media are now intelligent, ubiquitous, sensory, locative, calculating and even communicative, often all this as well as operating beneath the threshold of our perceptions. What's more, these media are engaged in systemic linkages that transform the environment and even intervene in biological operations. The concretizations that such technologies now undergo can be said to not just to create associated milieus with aspects of the physical world but also of the psychosocial and biological.

In the following I do not propose to provide a fleshed-out chrono-topology of this media environment. The task would be too extensive and it is transforming too rapidly anyway. I do want to undertake the more humble task of pointing to some areas where I think Simondon can be productive for thinking about how to negotiate this new terrain, first starting by looking at where we've already been.

BRITISH CULTURAL STUDIES

The tradition of British cultural studies has undoubtedly been the dominant school of thought for understanding media for the past several decades in the United Kingdom and continues to exert a significant influence. Although it isn't a tradition that Simondon engaged with directly, his work does offer a response to structuralism, which is influential in much of its work.

Although structuralism's grip is now rather feeble, it is worth briefly engaging with cultural studies given that it bequeathed a certain disdain for technological causality in media theorizing that continues to the present moment.

One of the foundational texts of the tradition is Stuart Hall's 'Encoding/Decoding', which famously destabilizes the linear Shannon-Weaver model of communication by pointing to the ways by which a message can be disrupted between its initial encoding and eventual decoding. Thus the message is open to a range of influences that will determine the nature of its eventual decoding. For example, such influences might include the visual 'languages' used in TV production, which relay coded meanings about the images used, or the context of its reception, whether this involves negotiation or opposition.

The upshot of Hall's essay is a shift from the concentration on media texts effects on individuals to understanding these texts as 'moments when the larger social and political structures within the culture are exposed for analysis' (Turner, 1990: 94).

There are two aspects of this I want to take issue with from a Simondonian perspective. First is the residual hylemorphism of this account and second is the notion of structure that emerges.

Regarding hylemorphism I agree with Matthew Fuller's reading of Hall's essay that despite being a useful text it fails to depart from the form/content model and as such it 'does not provide for a full account of potential media practices' (Fuller, 2005: 22). That is to say that the model still maintains the imposition of form on to content either in the process of encoding or decoding, a situation Fuller describes as cultural studies being 'trapped in receiver mode'.

The consequence of this is a focus on dominating structures as well as an inability for content to break free from its structuration by them. Such a model is therefore inadequate for fully expressing a situation in which 'media elements possess ontogenetic capacities as well as being constitutively embedded in particular contexts' (Fuller, 2005: 22). The model thus fails to account for the technicity of media, its mode of individuation and how this enables the modulation of cultural structures. In short it lacks a way by which it can describe the 'transformation of a structure into another structure' (Simondon, 2013: 529).

The main means by which the subject is understood as being structured in British cultural studies is through ideology. As Graeme Turner states, 'Ideology is the most important conceptual category in cultural studies' (Turner, 1990: 197).

It isn't necessary to go into detail here regarding the vacillations in the development of this concept, from its initial outline in Marx, and through its various avenues of development with Barthes, Levi-Strauss and Althusser. What is of interest is that Simondon's theory crucially gives a counter-theorization to the structuralism of the ideological approaches that have fed into British media and cultural studies. Although writing at a time when structuralism was highly influential, Simondon doesn't tend to engage with it directly but nonetheless it's hard to read him without sensing that he's aware that his own work pushes against the structuralist grain.

It's worth outlining here the manner of this difference in order to specify how a Simondonian media theory would find lacking those approaches that maintain a sustained focus on representation and discourse. At the heart of this difference are distinct disagreements regarding the related problems of the nature of the subject and its relation to the social (collective).

Like Simondon, Althusser opposes a hypostatized account of the human subject and part of his ideological approach is therefore to oppose humanism. As such he understands ideology as a profoundly unconscious 'system of representations' (Larrain, 1979: 155), which constitutes a structural aspect of any society and whose function 'is the cementing of its unity' (Larrain, 1979: 156).

In Althusser's theory ideology acts as a support for subjects to tolerate their situation through constituting them (through interpellation) 'in their imaginary relations to their world' (Larrain, 1979: 163). The concept is developed further by Gramsci with his theory of hegemony, which argues that the dominant ideology in a society shapes and controls the discourses by which that society understands itself, thus making it difficult to question let alone overturn.

In British cultural studies this general model is taken up by Stuart Hall to understand ideology as that by which media is connected to society. To a certain extent Hall maintains the unconscious operation of ideology by stipulating the 'invisibility of the process of signification' (Turner, 1990: 204), a process by which subjects experience the naturalness and 'common sense' of media messages.

At its heart cultural studies is concerned with the contestation of this common sense, that is, with challenging hegemony in the production of meaning. Having been developed following Saussurian linguistics, in which meaning is culturally attributed rather than in any way inherent in things, it was an easy step to adopt the Gramscian line regarding hegemony, which enables some sense of how some meanings become and remain dominant.

As might be recalled from the introduction, one of Simondon's aims was for an axiomatization of the human sciences in accordance with his ontogenetic account of individuation. One aspect of this project was to critique the inherent substantialism of existing psychology and sociology, the former for its focus on a rigidly structured, individualized and internalized subject, the latter for the over-determination of a substantive social field upon psychic reality. Simondon proposes that instead of the over-emphasizing of one particular pole of the social relation over another, what is required is a psychosocial account in which what have been taken as two separate terms are in reality co-constitutive.

One problem with structuralism of the type we find in Althusser and Gramsci is that it severely limits the nature of any such co-constitution as it predetermines its understanding of the subject as a consciousness that is always already structured through discourse and representation. As such it follows a rather hylemorphic model of the subject as that which is *subjected* to the determination of a dominant system of representations. What's more, such determination, operative at the level of discourse, is also responsible for the unifying structuration of the collective.

However, what is left unanswered is how either the subject or the collective come to be individuated, and indeed individualized. From this perspective the subject's development appears separated from nature and is purely cultural, through the operation of discourse. Even so this hardly gives an account of the subject's individuation.

Similarly, the genesis of the social is left unexplained. Instead it is assumed as a similarly hylemorphically formed substance moulded at the cultural and discursive level. That is to say that it appears as always already individuated via the hidden ideological structure.

We can then see that Hall's view that it is ideology that connects media to society ensures that the role of media is only ever operative as part of culture (discourse), it doesn't entail a causal role for technology *qua* natural causal object.

There are some clear lines of contrast to this structuralist and representational approach in Simondon's work that are helpful for sketching what a Simondonian approach to media would entail. These differences can themselves be aligned with the slightly different approach Simondon took in *L'individuation*, with its emphasis on allagmatics and individuation, and the approach adopted in *Imagination et invention*, which focused more on the relation of organism with milieu.

With its concern for the operation of the convertibility of structures allagmatics provides a clear juxtaposition to structuralist claims for a synchronic structure conditioning or constructing subjective consciousness. Notwithstanding the lack of any genetic account of the invention of this structure the ideological approach also lacks, as I've indicated, an adequate description of the individuation of the subject or the collective or their relation. For Simondon, as we will recall from chapter 4, the individuation of the subject involves a much wider and more complex set of operations than just ideological interpellation.

The ideological perspective on the subject lacks the topological and chronological structuration achieved via the dimensions of affect and emotion, nor that achieved via perception and action, which Simondon stipulates as necessary prior even to the achievement of the individualization of the subject. What's more, we must recall that this individualization is also the co-individuation with the collective, and that this always involves more than just discourse, but at the very least affect, norms, actions, emotion, belief, work, as well as representations.

As such, from the ontogenetic account of the development of the individual set out in *L'individuation* we can see that an account of the subject as ideologically determined is an inadequate account of the genesis of both the subject and the domain of ideology. From this perspective the passage of representations would seem to be more descriptive of one aspect of the inter-individual mode of the collective, lacking as it does the affective-emotive dimension coupled with action of the subject, which also enables the individuation of the transindividual. The transindividual relation is one that involves the concurrence of action with emotion, in which an action has a level of objective reality external to the individual, which also enables the emotive subject connection to the world that is also constitutive of its own self-reflection.

In *Imagination et invention* we can also find other resources for being sceptical of broad claims made for discursive approaches to media as well as the necessity for a broader approach for understanding media. In this text Simondon describes the development of the imagination as a four-phased progression from its genesis in the unstructured action and perception of simple organism through to its systemization and the creation of a mental model of the world, onwards to invention. There are several important implications of this account for the consideration of the cultural studies approach to media through ideology and discourse.

First, for Simondon, the symbolic must be understood as emerging from the development of the image, which itself is relatively autonomous to the subject. This entails the priority of the image with respect to language in that the image's role in the operation of signification and symbolism occurs after its initial biological anticipatory phase and during the latter stages of the cycle when experience becomes organized through perception and systemization. As such it is the image that develops prior to the development of symbolic functionality and that is also responsible for it. Language, as it were, arrives later on the scene and as a result of systemization during the image-cycle. This deflates the sails of a structuralism that sees language as the over-determining factor in the structuration of both the subject and the social. Indeed, the symbolic field is productive but needs to be understood as operating across a number of levels with which it is both determining and determined.

The use of symbolism to instigate a separation of culture from nature is resisted, along with a post-structuralism: meaning is not just produced through the arrangement of signs in relation to an underlying societal system of understanding. The system of signs isn't arbitrary. This is too narrow an explanation and also far too closed. The symbolic needs to be understood as ontogenetically developing through a real relation between the organism and its milieu and is only possible because of this. The symbolic emerges from the image and this is a structure that is intermediate between world and subject. That is to say it is both natural and social, if we want to use those terms. As Simondon describes it,

> the world of symbols is a species of pandemonium floating between situations of object and that of subject, interposing between the living and the milieu. (Simondon, 2008: 137)

As such the operation of symbols should not be reduced to just that of the discursive realm if this is understood as wholly cultural. What Simondon is proposing with his image-cycle is a system of recurrent causality 'that runs from the mental to the objective real through social processes of cumulative causation and also runs from the objective real to the mental' (Simondon, 2008: 13).

This is not to say that we should dismiss the importance of discourse and symbolism but it does need situating within a much broader system of causality that includes affect, tropism, emotion, memory, myths and crucially recognizes symbolism's role in the invention of image-objects such as the aesthetic and technical that themselves operate in the production and organization of the cultural at the level of both the individual and collective.

To assert that the symbolic is something that the subject is *subjected* to is too extreme. Rather the symbolic operates as one aspect of the organization and invention of the world; 'the symbol is a mix of subject and object that has instrumental value for invention' (Simondon, 2008: 138).

Obviously, we don't want the representational theories that emerged from structuralism to be set up as straw men here. It's certainly not my intention that these theories are portrayed as fully fledged idealisms; however, what they do attempt is a distancing of the cultural from causation that lies outside of human agency. That is to say that they insist in 'a difference in kind between cultural and natural events' (Lister et al., 2009: xv). As we saw in the last chapter this is an issue that interested Latour and is at the heart of his assertion of agency involving natural and technological actors.

Although some theorists have resisted the hegemony of discourse, such as Barthes, who proposes jouissance as a bodily and natural expression outside the scope of ideology, there is a narrowness to this resistance and it has resolutely failed to influence the orthodox position regarding technology in British media studies.

As is well documented[1] the denial of technological causality in favour of cultural agency was a result of Raymond Williams's trenchant dismissal of what he saw as Marshall McLuhan's technological determinism. Williams opposed this determinism with a cultural science, which was 'concerned with the necessary differentiation of its procedures from those of natural science' (Williams, 2004: 122), as well as the inclusion of *intention* into accounts of 'social and cultural process' (122).

As such he read McLuhan as promoting 'an ideological representation of technology as a cause' that was in contrast to cultural science's understanding of it as 'at once an intention and an effect of a particular social order' (Williams, 2004: 123). The result of this was to remove technical agency from cultural scientific accounts, a move that rings hollow with Williams's own claims about culture and ordinariness:

> What kind of life can it be, I wonder, to produce extraordinary fussiness, this extraordinary decision to call certain things culture and then separate them, as with a park wall, from ordinary people and ordinary work? (Williams, 2001: 12)

It is worth observing here that such a view necessarily must also attempt to purify culture from those influences it receives from technoscience in the shape of technical schemas it might use to understand reality. We can thereby see a similarity with Heidegger in Williams's grab for cultural purity.

Although renewed focus has been applied to the role of technology in theorizing media in recent years, mainly due to the prominence of the newer, digitally networked technologies, it's still the case that any influence they're allowed is culturally conditioned. For example, in one recent media and communication textbook called *Media, Culture and Society* (Hodkinson, 2011),[2] the author correctly dismisses a narrow account of technological determinism but then immediately extends this to a denial of any causal influence for technology that isn't culturally conditioned.

So although the author notes the pluri-functionality of some technologies, he is quick to point out that it is the human agent who chooses how they are used and that all technologies are the product of 'human needs, purposes and arrangements'. In this picture technology's role is always already subsumed to the cultural context so that the question can be posed: 'So how do we break down and study the significance of the contexts in which technologies circulate?'

The answer given is simply that we can categorize technologies regarding the capacities they have that have significance for culture. However, it is always the case that these capacities themselves 'are all dependent on human priorities and contexts'. Thus technologies have importance for the affordances they have but these affordances are always socially shaped. With technological causality thus tamed the humanities scholar is free to situate all the action within the traditional cultural arenas such as texts and discourse, ownership and identity. My intention here is not to denigrate this tradition of media studies. Far from it, much of this work is important and vital for comprehending the world in which we live. However, following Simondon, I do want to address the significant blind spot much of this work has regarding technology, especially given that our environments are so technologically laden.

FRIEDRICH KITTLER

Not all media theorists share this cultural bias regarding technological causality. Indeed, the most famous line of the most famous German media theorist, Friedrich Kittler, is, of course 'Media determine our situation' (Kittler, 1999: xxxix), a claim that requires some explaining.

If one aspect of German media theory is its resistance to the technophobia of critical theory (Siegert, 2013: 54) then this is one function that has yet to find purchase in the context of mainstream British media theory. This is still very much in the grip of either a liberal theorizing of the media's entanglement with the public sphere or, as we have seen, a Marxist-inspired discourse analysis and political economy.

For the purpose required here I'm interested in examining some of the broader themes that run through Kittler's work in order to help crystallize the position of a Simondon-inspired media theory. There are certainly some areas of shared interest between the two thinkers such as a resistance to anthropologism, the claim for an autonomous technological evolution and a championing of the engineer. There is also a shared interest in a technical *a priori*, or as Mark Hansen describes it, 'the contamination of thinking by technics' (Hansen, 2006: 298).

Both Kittler and British cultural studies share a structuralist heritage despite differing regarding the nature of this inheritance. Where cultural studies maintained the determinative power of the symbolic in structuring the subject whatever media types are ubiquitous, Kittler identifies the symbolic universe of language as that which dominates only during the period of the Gutenberg Galaxy. This specificity is achieved through utilizing Lacan's distinction between the symbolic and the real in order to account for how changes in the way that data is processed and transported are consequential for the nature of the subject.

Kittler calls the social wide conglomeration of techniques, tools and institutions of data production, storage and processing a *discourse network*. That's to say, for him, a culture is defined by the way that data flows through its discourse network. Kittler argues that the modern subject emerged as a result of the symbolic activities of speaking, writing and reading being the dominant way that data was communicated, recorded and produced during the nineteenth century. As such, and drawing from McLuhan, Kittler emphasizes the importance of the materiality of the communication process.

Thus, in his first book, Kittler describes what he calls the discourse network 1800, which is an account of how the discourse network that was constituted by the social practices and technologies of language use led to the constitution of the modern subject imbued with the enlightenment spirit. As such Kittler is in broad agreement with structuralism's claim that language is responsible for structuring the subject. Kittler's key move is in tying changes in the discourse network to parallel changes in subjectivity.

With the introduction of analogue electronic media there is thus a transformation of the dominant discourse network. To return to Kittler's use of Lacan, in which the discourse network of 1800 operated in the symbolic universe this new discourse network of analogue technological media enabled the recording, storing and distribution of data that is extra-symbolic, that is of the *real*. Where the discourse network of 1800 dealt with data flows that 'had to pass through the bottleneck of the signifier' (Kittler, 1999: 4), the newer analogue media could record and process the *real*, by which Kittler means the *physical*, which cannot be so passed.

Such a shift has obvious implications for the kind of media analysis performed by cultural studies in that the attachment to discourse and representation is now understood as an analysis that is stuck in the constraining universe of the symbolic. As Kittler claims, when discussing Foucault, 'Discourse analysis cannot be applied to sound archives or towers of film rolls' (Kittler, 1999: 5). Quite simply, the data that analogue media use overflows the bounds of the symbolic.

The transformation of the dominant discourse network 1800 and the data it handles is responsible for the shift to a new discourse network 1900. Kittler links the changing of discourse networks to Foucault's notion of epochal shifts, which are caused by the transformation of the dominant *episteme* (the practices and theories that condition the possibility and organization of knowledge). Where Foucault fails to offer a convincing account of the dynamic of these cultural breaks Kittler directly connects the transformation of the discourse network with the overturning of forms of knowledge. This change in the kind of data being produced and communicated on the networks, from symbolic to extra-symbolic, means the previous symbolic discourse networks loses its monopoly and power.

Thus for Kittler, Foucault's descriptions of a discourse analysis based on the *archive* remains rooted in the symbolic universe associated with language and print-based media. As such it is unable to adequately address the kind of data flows that operate with analogue media as its descriptions remain symbolic while the newer media are concerned with *signals* and the processing of data that is not necessarily semiotically analysable. So although Kittler's structuralism coincides to that of British cultural studies, it only does so up to a point, which is during the period of discourse network 1800. Beyond this point its focus on the symbolic universe is too narrow to grasp the kind of data that analogue media are producing and circulating.

Like Simondon, then, Kittler is specifying a direct link between technological and human transformation and as such also has in his sights the notion of humanism. Unlike Simondon, Kittler is not seeking a humanism renewed through a modification of the technical relation. Rather, Kittler proposes a technical *a priori*, which not only has established Man, the enlightenment subject, through the operation of a certain discourse network, but by changing will be responsible for his eradication:

> So-called Man is not determined by attributes which philosophers confer on or suggest to people in order that they may better understand themselves; rather, He is determined by technical standards. Presumably then, every psychology or anthropology only subsequently spells out which functions of the general data processing are controlled by machines, that is, implemented in the real. (Kittler, 1997: 133)

What is responsible for this disappearance of 'so-called Man' is the further epochal shift from analogue media to computing. Where analogue media such as gramophones and televisions record and process the effects of the physical real—sound and light—with computing there is a shift beyond a form of mediation that is correlated with the human sensorium. That is to say that with Shannon's theory of information what is attained is an overcoming of the distinction between media and the physical distinctions with which they operate (light waves, sound waves, etc.) as all of these are subsumed by information, which is granular and mathematically processable.

It is debatable whether Kittler actually proposes an informational ontology 'in which only that which can be switched exists at all' (Kramer, 2006: 105) or if he is just restricting his description to the operation of communication in discourse networks. As Kramer points out, if it is the former then 'everything that can be switched is essentially invisible to the human senses' (2006: 106) and this is tantamount to claiming that human phenomenal experience is an illusory epiphenomena of a reality that is beyond human sensibility. If Kittler isn't making such a grand ontological claim it's still the case that the operation of human communication and meaning is now determined at a technological level that is beyond human sensibility. Thus with digital technology it is not so much the case that the conditions for knowledge are determined by a discourse network but that knowledge disappears as the conditions for it are ungraspable: ‘the machine substitutes man as the referent of communication’ (Kramer, 2006: 107).

Whichever interpretation is taken Man turns out to be determined by media.

It is worth pausing here to consider how Simondon can be contrasted with Kittler. It’s not surprising that there are some commonalities between them given Kittler’s influence by Foucault, who was also working in the same tradition of French scientific epistemology as Simondon. Like Simondon, Foucault was deeply influenced by Canguilhem and it is therefore no surprise that they both shared an interest in the notion of epistemic breaks.

As we’ve seen, Simondon also tied his notion of epistemic development to technology. This is most obvious in his work on technical mentality in which the operation of technical modalities is reflected in the operational schemas of knowledge. This is also complemented by the theory of the image-cycle given in *Imagination et invention* in which technical objects, among others, are components in the processual development of the imagination and ideas.

The two thinkers complement each other clearly in this respect and it is easy to read Kittler’s account of the co-development of research into cognition and perception with advances of media alongside Simondon’s work on technical schemas.

What can’t be reconciled are the ontological commitments the two thinkers make that come to a head around the notion of information. Whether Kittler ultimately holds an informational ontology or not it is clearly the case that the notion of data is prioritized for him. A culture can be defined by its discourse network, which is basically the way it deals with data, and as we’ve seen the kind of data a culture deals with changes with media technology.

If Kittler holds an informational ontology it is a fundamentally different one to Simondon, given its reliance on Shannon’s theory of information. Simondon’s notion of information departs radically from this and this difference has significant consequences. The most obvious of these is that Simondon’s theory is non-reductive: he’s not interested in reducing everything, or even all media, down to mathematical information. For him information is productive of new levels of reality. Rather than seeking a way to subsume all media to digital data, a Simondonian theory must take into account the way media are involved in the development of new structures and how they bring different orders of magnitude into relation.

Networked digital media cannot, on this account, be understood as the Ur medium that eradicates media and Man, leaving just the endless circulation of data. Instead they need to be understood as creative of new unities, new structures and new relations.

Although Simondon may have resisted anthropologism, this does not mean he desires the disappearance of Man. Although he agrees with Kittler that technology and humans are co-constitutive, it is by no means the case that the disappearance of the human is foreclosed in this relationship. That's not to say that it might not be one day. Simondon's thought could easily be used to argue for the likely evolution of the post-human. However, for the moment, this isn't what concerns us.

What is of concern is the question of whether these new technologies present a danger that requires the formulation of a new humanism, just as Simondon claimed was required during the industrial era. One thing Simondon teaches us is that we need to work to grasp the consequences of our relation with technology to make it adequate for both human and machine.

I think Kittler is useful in helping us formulate some of what is at stake here. His descriptions of the various discourse networks help us understand the epistemic implications of technological change, although his example needs to be challenged given his 'exclusion of the body as a medium' (Kramer, 2006: 95). Simondon is much clearer regarding the broader biological and psychosocial individuations in which technology is implicated. Thus for him it simply cannot be the case that 'everything that can be described, can be represented in the terminology of technological processes' (Kittler as quoted in Kramer, 2006: 95).

But what Kittler also adds is an understanding that technology can operate beneath the threshold of the sensible and cognitive. That is to say that these media overflow the human capacity of awareness:

> The coupling of a storage medium and a transmission medium, of a typewriter and a radio network, finally resulted in a universal medium of computation, that is, a machine capable of registering, transmitting, and computing any data whatsoever without human intervention. As Turing has proved, the computerized calculation of recursive functions really exhausts the whole domain of computability. (Kittler, 1997: 126)

Rather than heralding the end of 'so-called Man' what the realization of this understanding should do is indicate the radical environmentally of some of our media in relation to us. Given that Simondon's solution to the alienation of man with technology was to ensure their mutual co-operation, what is the solution when the technology operates beyond any possible real-time relation with the human sensorium? Does this mean that we should now give up all humanism and accept that we are now entirely conditioned by a technological *a priori*? I will turn to this in the next section.

Another similarity between Kittler and Simondon is that both thinkers reserve special admiration for engineers and inventors. Kittler is especially praiseworthy of Edison, who is responsible for two of his great technical triumvirate—the cinema and phonograph—which he argues are responsible for the transformation to discourse network 1900, along with the typewriter.

Kittler was himself a computer programmer and advocated the teaching of programming to students. There is a certain perversity in this given the purported impotence of human intentionality in the face of the informational *a priori*, but as Winthrop-Young claims:

> And while this results neither in Western Enlightenment nor in any mastery over digital machines, it will at least enable us to rise above our self-caused software-supported immaturity and interact eye to eye (or signal to signal) with all that is on the verge of leaving us behind. (Winthrop-Young 2011: 77)

Simondon's argument for a technical pedagogy is rather more coherent given the role it would play in bringing about the truly technological culture that he advocates, and thus is a more positive activity than just being a mark of curiosity regarding the nature of that which is about to consume us.

The final aspect of Kittler's work worth mentioning is his notion of the autonomous evolution of technical media. Given that the (so-called) social, such as it exists for Kittler, is determined by the technological, it seems reasonable to ask how this technological dynamic operates. It is clear that with Kittler we have found the diametric opposite of the culturalism of Raymond Williams: 'To begin with, one should attempt to abandon the usual practice of conceiving of power as a function of so-called society, and, conversely, attempt to construct sociology from the chip's architectures' (Kittler, 1997: 162).

And neither is the answer to be found in a human-centred instrumentalism: 'Technical media don't arise out of human needs, as their current interpretation in terms of bodily prostheses has it, they follow each other in a rhythm of escalating strategic answers' (Kittler, 1997: 121).

Although Kittler gives some convincing accounts of the ways in which technical developments help give answers to various problems, they are often not convincing as an account of the autonomy of the dynamic given that they usually involve some human figure (an engineer or soldier of some rank) and/or are driven by the necessity of the military-industrial complex, which is still difficult (although not impossible[3]) to conceive as not involving human intention. However, Kittler does his best to insist that technical media are not cultural but are 'strategies of the Real' (Kittler, 1997: 129).

This description actually makes more sense from a Simondonian position given that Kittler evidently doesn't wish to allow human intention into this reality, given its qualia-like status. However, the development of one form of media does prefigure, or at least provide the potential for, another. For Simondon, we may recall that the dynamic of technical evolution occurs through progressive concretization, which is quasi-autonomous from human intervention. Human inventors are the handmaidens of technical development. If we're generous we could also allow Kittler this interpretation.

What is significant as far as thinking about Simondon and media is concerned is to what extent are psychosocial and cultural aspects intimately involved in these concretizations? It is clear that Simondon distances technological development from contamination by the cultural. I've already mentioned his horror at the decorative aspects of modern automobiles that defile their technicity. What then becomes of technicity when its operation relies on the psychosocial?

MEDIA AS ENVIRONMENT

So far in this chapter I have looked at two approaches to theorizing media. The first, British cultural studies, is too restricted in its focus on discourse and representation, so it fails to take into account that which overflows this purview. That is, it fails to account for the pre-individual milieu in which discourse comes to be individuated. The work of Friedrich Kittler addresses this concern by specifying a technical *a priori* responsible for structuring possible discourses, but his notion of data and information was found to be too reductive.

An interesting aspect of both of these theories is how they understand technical media as, to some extent, environmental. We may recall Hodkinson's description of technical media in terms of the affordances they possess restricting the type of communication that can occur through them. As such this description reduced technical media to the environmental constraints they imposed on messaging.

Kittler, on the other hand, understood media in a more infrastructural capacity in regards to how they enable the production, storage and processing of data. As such their role is as a determinant structuring culture.

As we have seen, Simondon's interest lies in the area between determinism and indeterminism, thus we need to be wary of warranting the cultural or technological a position of over-determination. When thinking media with Simondon we can have sympathy with Latour's suggestion to think in terms of relations; however, what Simondon gives us is a richer ontology to think about what the actors are and how they come into relation to create unities of operation. Significantly, in doing this, he also helps us think through the implications of the openness of systems and to always keep in mind the indeterminism this involves as well as the relationality of any relations that are individuated.

As such Simondon shares Latour's impulse to think of media as environmental in the strong sense of that which not only supports but also involves feedback effects on the human, the world and itself. To think media in this way is complex. This is the challenge of Simondon's encyclopaedism. To attempt this requires the ability to focus on one order of magnitude or one operational level, while keeping in mind its relation to others. It requires a commitment to the interdisciplinary—and with this Simondon was also ahead of his time—and ontogenetic thought.

What I wish to do in the final sections of this chapter is attempt to use Simondon to work to describe the outline of an approach to theorizing media. Although Simondon describes a narrative of the relation of culture and technology, of historical progress, of a relation of the human to the world, of co-evolution, he does not offer us closure. There is no unhappy ending, as with Kittler. Instead there are ongoing problems of relation, the development of new technological and natural forces in which humanity must develop a unifying and mediating role. If we are to reduce technical media to just constraining our communication via affordances, then we do them a disservice. There are far greater risks and problems associated with them, as well as greater rewards and opportunities for psychic and collective development.

Stiegler has correctly identified this in his description of contemporary digital networked media as pharmakon, although his analysis is constrained by phenomenological restriction of the technical to the mnemic. What Simondon offers is an approach that considers a broader range of concerns implicated in technogenesis across varied levels.

Recently there has been an increased interest in the role of technical infrastructure for comprehending the current condition of mediality. As we've seen, Kittler's project was itself premised on the effect that technical infrastructural change had for culture and knowledge. In opposition to Kittler's reductionism vis-à-vis information, with Simondon we propose an ontogenetic realism of unities of operation, levels and orders of magnitude.

One of the problems with considering contemporary media is the sheer complexity of the chrono-topological structures we are dealing with. It is inadequate to just try and account for these via humanistic perspectives such as instrumentalism or discourse. These fail to do justice both to technics and to the unities they individuate with. It is clear that to think with contemporary media technology what is required is a comprehension of the stratification they involve.

Notwithstanding the magnitude and range of scales of hardware on which the manifold services of the Internet operates (from microtransistors and fibreoptic cables, satellites and servers to routers and devices of varied sizes) there is also the limitless range of software that can be instantiated on Turing machines that both shrink and multiply around us at a dizzying pace. Additionally the manifold levels at which these interact with cultural and natural entities also requires articulation. With contemporary media technology there is the immanent and continual development of infrastructures for the composition of new unities of operation that can and often do involve biological and psychosocial individuals of varying orders of magnitude.

What I have in mind here is not just empirical engagement with figural technical objects but the creation of new unties in informational connection with one another. In a sense we are engaged with a new reticular structure in which the network is composed of imbricated levels and keeps shifting its key points. As such we need to be able to describe the stabilities and instabilities, the phasings and dephasings bringing forth and being structured by modulation and regulation.

And it is not just technics that requires such an analysis: work needs to be done on rethinking the notion of the subject; for example, what is the nature of individualization that involves the dispersal of subjectivity in technical processes that operate below our sense of perception or involve so-called intelligent processes?

Such work must also engage with the nature of operative unities and their openness to relations and disparities they may encounter. It would also engage with the epistemological and normative consequences of these structures and the schemas by which we understand them.

The logic of this proposal should follow from its roots in Simondon's work and is therefore also unashamedly encyclopaedic in nature as well as open to being situated between the empirical and transcendental, and between determinism and indeterminism, where complexity is situated.

At the very least Simondon offers us an ontology and epistemology that allows us to situate empirical work on media, but more importantly he gives us the tools to evaluate both our media technological use as well as the fitness of schemas that have been developed from them to understand the world (I will discuss Big Data as an example shortly).

Simondon asserts that to understand technical mentality requires understanding subsets as detachable from the whole that should be studied 'considering it in its entelechy, and not in its inactivity or its static state', that is, it needs to be studied at its level and inductively, as a regime of operation.

Although writing in the very early days of the development of the Internet, Simondon envisioned the shift in the technical mentality that was occurring from the industrial to the post-industrial networks. There are a number of important aspects to this development.

As we've seen, the operation of industrial technology led to a situation of alienation, which was ultimately due to the constraining of worker's relations with technology through specialized roles, which led to these relations only ever being partial. The managers and workers all experience only certain aspects of technical reality and thus 'they cannot elaborate a value code that is capable of becoming universal'. With post-industrial technology, which is more properly reticular in structure, Simondon sees the potential for a situation in which the human-technology relation is more universal, as the fragmentation is reduced 'the distance between the inventor, the constructor and the operator is reduced; the three types converge towards the image of the technician'.

This is made possible because after the development from artisanal technology, where the energetic and informational relations with technology are provided at the human scale, to industrial technology whose gigantism led to a situation in which the modulation of energetic inputs from vast natural reserves involved fragmentary informational interventions, there is a return to technology where the energetic and informational scale is again that of the human operator of electronic devices.

However, for this to occur Simondon recognizes that there is a requirement for a stable network infrastructure to be in place upon which inventive and open technologies can be continually developed. As such Simondon recognizes the 'post-industrial' object as 'the unity of two layers of reality'; there is the permanent infrastructure that is maintained by the ability to mass produce components that operate at required standards and upon this there are the personal, more temporary devices that require the infrastructure to operate (for example, cars on a road network, smart phones on a data network).

> The [technical] object is not only structure but also regime. And the normalization of thresholds of functioning expresses itself in the difference between relatively separate subsets [of the whole]; the degree of solidarity is precisely the measure (in the Greek sense of 'metrion') of the relation between the permanent parts and the parts subject to replacement. This measure is what defines the optimum of the regime in the relation of thresholds of functioning. (Simondon, 2009: 24)

The technical object needs to be understood as the operative conjunction of both infrastructure and devices that together work in what he calls a regime. The use of the Greek term *metrion* is important as with it Simondon is stipulating that he is not referring to strictly physical measurements (for example, weight, length, etc.) regarding margin for operation of the devices on the network infrastructure but to *due measure* that involves the application of the kind of judgement that the technician, designer and artist would have in judging the appropriateness of operative activity both normatively, affectively and aesthetically.

What Simondon is aiming for is a technical development that achieves a unity of the operation and development of cognitive schemas, affective modalities and norms of action through the maintenance of their openness for further development and further amplification. As such the contemporary

Internet, at first glance, would seem to fulfil these criteria, perhaps even going beyond them to provide a modulating infrastructure of platforms.

It is my contention that Simondon provides a way for us to consider the kind of networked reality that helps structure the contemporary world. He provides an ontology and method that is designed to think though the causal operation of unities at different orders of magnitude. These are unities that involve, as we've seen, the inter-relation of those realities from the varied regimes (physical, vital, psychosocial) through the technical. What's more, Simondon also supplies the means by which this ontology is also at the same time an epistemology. Importantly, too, with his theory of technical mentality he stipulates that if this technology and our culture are going to develop beneficially then we require *due measure* to balance the affective, normative, aesthetic and cognitive with those technical developments. In what follows I wish to think through some aspects of contemporary environmental media technology using this as a guide.

BIG DATA

Utilizing a technical schema to explain social organization isn't new and finds one contemporary expression in the notion of Big Data. With a keen interest in not just mapping the social but in extending its functionality to commanding and controlling it, it can be understood to be in the same tradition of socio-cybernetics as propounded by Wiener and Beer, who proposed the application of cybernetic methods developed in engineering contexts for social regulation.

Essentially, the method of Big Data is little more than an invigorated empiricism and is demarcated by the three Vs: the collection of huge volumes of data at real-time speed (velocity) from a wide range of sources (variety). The strong claim made for this method is that the data will be so rich that patterns will arise spontaneously from it, revealing an underlying sense. Chris Anderson (2008) evangelizes it thus:

> This is a world where massive amounts of data and applied mathematics replace every other tool that might be brought to bear. Out with every theory of human behaviour, from linguistics to sociology. Forget taxonomy, ontology and psychology. . . . We can throw the numbers into the biggest computing clusters the world has ever seen and let statistical algorithms find patterns where science cannot.

Academic proponents of Big Data, such as MIT's Alex Pentland, also make impressive sociological claims for it:

> Adam Smith and Karl Marx were wrong, or at least had only half the answers. Why? Because they talked about markets and classes, but those are aggregates. They're averages. . . . This is the first time in human history that we have the ability to see enough about ourselves that we can hope to actually build social systems that work qualitatively better than the systems we've always had. That's a remarkable change. (Pentland, 2012)

However, if we investigate Pentland's theorizing of the constitution of the social (as described in his book *Social Physics* [2014]) we find that rather than arising from the data itself, it rests on an analogy with the operation of online social networks such that for him an improved social system 'might look a lot like Wikipedia but founded on overlapping clusters of buddies who have face-to-face relationships' (Pentland, 2014: 209). What is key is how ideas flow through social networks and in doing so shape 'the norms, productivity and creative output of our companies, cities and societies' (Pentland, 2014: 4).

Such an ontology entails that individual's own motives are subsumed to *overall social network effects* to the extent that Pentland claims that 'ideas flow is the real story of community and culture. The rest is just surface appearance and illusion' (Pentland, 2014: 44).

Pentland's ontology of the social is interesting in that it can be understood as a hybrid between a simplified semiotics (now instantiated in the notion of 'ideas flow') with the technical schema of the network. Additionally, Pentland also adds a very cybernetic concern for behaviour and purpose when he also argues, in anthropological mode, that humans are habitual creatures and can thus be observed 'in just the same way we observe apes or bees and derive rules of behaviour, reaction and learning' (Pentland, 2014: 190).

From the above it should immediately be apparent that in stipulating his rather limited social ontology Pentland is himself imposing a theoretical *a priori* upon any data from which an interpretation should emerge. What's more, the ontology he has developed is merely a mirror of that most contemporary of technical schemas, the social network. Simondon's warning regarding analogically applying technical schemas to inappropriate domains is prescient, that although a schema may describe certain causal aspects of a domain it won't 'exhaust' it. In this case Pentland has mistaken the basic effects of the social network schema (ideas flow) as accounting for the whole of the social domain. As such what Pentland is guilty of, in Simondon's allagmatic terminology, is mistaking the resemblance of a structural relationship for the true analogy of 'identities of operative relations' (Simondon, 2013: 533).

Something else that Pentland shares with the socio-cybernetics of Wiener is a focus on understanding social systems in order to maintain them in a state of homeostasis or 'resilience'. From a Simondonian perspective this is both naive and dangerous. As should be apparent it is simply not the case that the social can be understood as a system that is closed enough to maintain a stable equilibrium. At best, due to its necessary openness, as well as the different types of recursive feedback loops that cross it and help constitute it, the best that can be hoped for is a metastable equilibrium, which is capable of a coherent individuation of phased development. An understanding of the role of a socio-cybernetics as maintaining a stable social equilibrium would be to impose it hylemorphically, which would risk systems that are incapable of being truly inventive but instead have as their aim a single purpose. Pentland's mistake, reflective of the problem of the bad analogy he makes, is that he tries to make a theory that may be of some use in controlling relatively closed structures work for the radically open system that is society.

Pentland is working with a probabilistic account of information, which he contends will enable the moulding of behaviour via measurement and management. As such it is clearly a cybernetic theory. However, as Simondon makes clear with his reconceptualization of information, this theorizing of information elides the indeterminism at the heart of being. Instead, what we find in Pentland's account of Big Data is the combination of two technical schemas, the cybernetic and social network, neither of which is adequate for the reality they are being used to explain.[4]

SOFTWARE

One requirement for developing a Simondon inspired theory of media technology would be to identify the various levels at which contemporary media operate in order to then develop a relational account. For example, we can describe the various components and unities by which the infrastructure of contemporary media is composed such as cables, satellites, server farms and energy sources as well as the devices whose operations rely on them. There is also, of course, the software components that operate at both of these levels and that are responsible for platform structures that are involved in processes of collective individuation at various scales and of differing modes.[5]

Software has proven itself to be a rich subject for analysis and has not allowed for any easy definition. A range of theories have been developed to understand this recent technological development, from those who deny it actually exists (Kittler, 1997: 147–55) to those who believe we can better understand reality through it, as some proponents of Big Data are prone to do. Others have written about how software can be defined (Mackenzie, 2006; Fuller, 2008; Galloway, 2004), for example, regarding its materiality as code (Hayles, 2002).

Although it's not a subject that Simondon considered I think his work can provide some useful approaches to engaging with software. I will briefly propose two ways here, both of which involve understanding software allagmatically, that is, in terms of its operations and structures. The first considers some operational unities that involve software and the second considers some contemporary issues concerning the mode of individuation by which software develops.

Unlike Kittler, we do hold that software exists and has real effects, although there is not room here to engage with the knottier questions regarding the status of software per se. Such questions are beyond the scope of this text given that they require an investigation into the heterogeneity of ontic situations in which software are realized, leading to questions of being able to differentiate software from service, for example, or to unpicking the relation of software from hardware development, given that there is often a close relationship between the two in which their mutual delineation can be obscure.

Following Simondon I would like to suggest the reality of software as technical object in that it acts as an informational relation between two or more orders of magnitude, that is, to also say that software always involves a process of mediation. What these different orders of magnitude are, or to put it another way, what relation of which levels software is expressing, is dependent on the particular situation under investigation, and cannot be universalized. This is a function of the increasingly diverse situations in which computational technologies are found, a diversity we can partially explain as the coming into relation of two open and inexhaustible domains, that of computation with that of the pre-individual.

In this section I will attempt to answer some general questions regarding software and will do so by considering an aspect of its relation to the economic, a domain largely ignored by Simondon. Given the sheer scale of the subject matter at hand I offer here a few suggestions of possible areas for investigation, that is, as a sketch of how Simondon's work can help us develop accounts of software.

We will use as an example Urs Bruegger's and Karin Knorr Cetina's discussion of the networked operation of the foreign exchange market, a system that is an example of techno-social concretization underpinned by software (2002a; 2002b).

This market is composed of and maintained by the global interactions (via a multitude of networked devices) of traders using a software-based system that records, structures and displays these interactions back to the traders:

> Like an array of crystals acting as lenses that collect light, focusing it on one point, the systems collect and focus activities, interests, and events on the surface of computer screens. The screens themselves are identically replicated in all connected institutions and trading floors, forming, as it were, one huge compound mirroring device and site. (Bruegger and Cetina, 2002a)

It is the relationship that the traders have with the market as it appears on the screen that interests the authors. The screen is not just a medium to receive information but 'a building site on which a whole economic and epistemological world is erected' (2002a: 395). The traders interact with this world as if with a living organism. Before the introduction of networked computing the market had been comparatively uncoordinated and temporally disjointed, operating using telex and telephones. At that time there was a tension between the different information held in and about different localities within the global reach of the market. With the development of a single global computerized market system there is an overcoming of this local-global disparity.

> After the introduction of screens, the market became fully available and identified as a separate entity in its own right for the first time—with prices, interests and the relevant information all visually indicated on screen. The market on screen is a 'whole' market and a global presence; it subdivides into different information feeds and dealing systems, but these are configured to form a global picture framed by the boundaries of the screen, which also serves as a medium for transactions. (Bruegger and Cetina, 2002b)

The resolution of this disparity can be understood in Simondon's terms as a concretization. In this case the technical object that has been invented is the market system, which is the regime of devices and instantiated software as well as traders interacting with it. Without the traders the market-as-individual would not exist because its associated milieu (the market world) would not be operational. What makes the associated milieu of this particular individual significant is that it is constituted by activity from the psychosocial regime and not, as with Simondon's usual technical examples, the physical regime. Although we can imagine the network operating without any trader interactions, it is only with these interactions that the system operates fully as a system and original virtualities are uncovered. It is, after all, the involvement of the traders that is a condition for the system's continued operation.

It's clear that an extraordinary amount of technical infrastructure needs to be in place for the market system to exist (for example, global networks, server farms, computing hardware and software). However, this form of technology also requires engagement from the psychosocial in order to generate an operative associated milieu and become truly technological. This does not just mean economic, political and institutional structures but also the affective engagement of the traders themselves. This kind of engagement is demonstrated by one trader who, when asked what the market was for them, responded:

> Everything. Everything. How loudly he's screaming, how excited he gets, who's selling, who's buying, where, which centre, what central banks are

> doing, what the large funds are doing, what the press is saying, what's happening to the CDU, what the Malaysian prime minister is saying, it's everything—everything all the time. (Bruegger and Cetina, 2002b)

One aspect of this technology, then, is that the cognitive and affective reactions of those who use it become part of its immediate operational structure. That such software-based technology operates in conjunction with an associated milieu constituted from the regime of the psychosocial indicates a change in degree from industrial technology. That is to say that we can understand that our everyday world, infused as it is with these kinds of networked systems, has become more decidedly techno-cultural.[6]

This infusion of the psychosocial with technical operation presents us with one point of indeterminacy where the epistemological project of Big Data flails, that is, its lack of ability to measure and predict the affective aspects of transindividual operations.

In chapter 4 we saw that affect is foundational for living beings and is also the subconscious grounding for psychism. That is to say that the temporal unity that affect is for a living being is always prior to rational thought and forms the basis for action. As the descriptions of the traders above indicate, their engagements are not just driven by rational calculation but have an important affective dimension as well.

It is nothing new to point to the complexity of the economy (indeed the idea of the hidden hand is a metaphor for this operational complexity); however, the claim made for Big Data that it can get to grips with this complexity misses the point that it's not just that the complexity is one of the scale of the number of interactions that make it undecipherable, but the nature of those interactions. Pentland's model of ideas flowing through networks as a schema for understanding the social is just as inadequate as the original cybernetic identification of animals with machines. It cannot adequately account for indeterminism from the environment, nor can it account for the indeterminism present in the affective core of the individuals making choices, nor the affective metastability of the transindividual relations they might enter into.

In his important work *The Power at the End of the Economy*, Brian Massumi is partially inspired by Simondon's work on affect and transindividuality, which he uses to argue against the contemporary neoliberal paradigm of the self-interested, rational individual. Like Simondon he prioritizes the affective, which rather than being the opposite of rationality, is actually the precondition for rationality reaching any decision to act. Without affect, rational calculation could extend its deliberations regarding a decision indefinitely, and may even lack the motive to make a decision. As Massumi puts it, 'Ratiocination chews its cud. Affect cuts to the quick' (Massumi, 2015: 48).

As well as offering a critique of self-interested rationality as a basis for understanding social and economic operation Massumi's work is vitally important in fleshing out an affective politics in line with that which is only hinted at by Simondon.

Alberto Toscano (2012: 94) correctly states 'we must acknowledge that an explicit thinking of the political subject is absent in Simondon'; however, the subject does remain a 'necessary but not sufficient condition necessary for any political activity' (2012: 95). That is to say that the way that Simondon theorizes the subject, although not explicitly political, due to its requirement for collective individuation made necessary by its inability to resolve the problem of its embodied immanence, means it will enter a situation that is nascently political.

We have claimed that a Simondonian politics would involve a restructuring of the social in part through a transductive amplification of values throughout the transindividual domain. In opposition to neoliberal political doctrine Massumi suggests an affective politics consistent with this Simondonian outlook in that it proposes modulating bodies in a state of affective *attunement*, which are primed for the transductive cue of a singularity that enters that metastable situation. Massumi theorizes the subject in a similar way to Simondon, that is, through a process of individualization, which is both psychic and somatic and involves both a relation to itself and that which exceeds it. As for Simondon, such theorization allows for the possibility of transindividual individuations that occur through the linking of polarities, which can also involve the activation of capacities. For Massumi this affective politics is also an aesthetic politics, which requires the maintenance of disparate tendencies in a metastable situation in order to create an environment ripe for a singularity to trigger an event with transductive repercussions involving transindividual structuration. As such affective politics requires the preparation for the possibility of an event that through *sympathy* might lead to an affective contagion: 'The communication of affection at the heart of the sympathetic event is by transindividual nature *nonlocal.* It can reverberate across the relational field, faster than the speed of conscious calculation' (Massumi, 2015: 84).

The results of this reverberation are by no means certain; the metastable situation it disrupts could de-individuate into fragments or lead to a further coherent individuation through an event of invention. Massumi restricts himself to using some recent political events such as the Arab Spring to illustrate these points and calls for a 'proliferation of practices aimed at developing techniques of relation' (Massumi, 2015: 94) in order to experiment with affective politics.

This could involve the practice of structuring affect through computational means. What we witness with software systems such as the financial market is the co-individuation of a technical system with a psychosocial individuation of the traders. The software-hardware (technological aspect) is causally efficacious in amplifying a transindividual modulation of the market traders, both in behaviour but also in affective relation with the market itself.

The recent work of Mark Hansen also draws on Simondon when considering the import of the 'the ineliminable temporal gap separating the operation of a technically-distributed system-environment hybrid from any subsequent cognitive or perceptual account of its operation in consciousness' (Hansen, 2012: 43).

Following Simondon's stipulation of the necessarily open nature of systems (in contrast to informationally closed autopoietic systems) Hansen develops the idea of the system-environment hybrid that reflects the 'double relationship' that any system necessarily has with both its associated milieu as well the pre-individual environment. What 'operational blindness' describes is the way that digital technologies can operate in our environment in such a way that the nature of their operations are inaccessible (perceptually or cognitively) until a later time.

Hansen uses the example of Étienne-Jules Marey's chronophotographic images that 'give us data about our perceptual processes' but because they are 'temporally distanced from the operationality that the data measures, this data can never obtain the status of lived experience' (Hansen, 2012: 43).

Returning to our example of the computerized financial markets we also witness the ever-increasing speed of data networks as well as the use of automated algorithms that undertake trading at speeds[7] far higher than human cognition can register. For example, in his article on what has become known as high-frequency trading, Toscano writes:

> We thus confront a compression of market-making transactions to speeds far below the threshold of individual human cognition, and an asymptotic acceleration of market turnover. The fastest trading chip executes a transaction in 740 nanoseconds (or 0.00074 milliseconds) while human reaction time to a visual stimulus is around 190 milliseconds. (Toscano, 2013)

Hansen identifies a specific danger, which is the 'engineering of the pre-individual' of our sensibility prior to the constitution of the phenomenological subject or 'bodily self-perception'. That is to say that such technical operations exert an indirect effect on conscious experience by impacting upon pre-conscious sensibility.

Thus we now have situations in which there is a *feeding-forward* by media technology of experiential data, which would otherwise be inaccessible to the human, to a later time than their capture. As Hansen says of this operation, in a Simondonian tenor, 'the "meaning" of the experience is "proper" neither to the human or machine, and so can't be understood exclusively from one or other perspective' (Hansen, 2015: 51).

What is described is an operational unity between the technical and human experience, a co-operation with the danger of a new kind of alienation in which the technical introduces new experiences for the human while simultaneously operating beyond its experience. Hansen is alive to the possible consequences of how such data could be used by contemporary capitalism, not least by those, like Pentland, who suggest the use of Big Data for socio-cybernetic control.

His suggested response also has a Simondonian tenor in that he proposes that this new technical-human unity requires regulating such that the data it produces is not available for instrumental manipulation but 'can be integrated into larger behavioural assemblages that will help us form our higher-order behaviour by modulating how media shape the sensibility from which such behaviour emerges' (Hansen, 2015: 198).

The point being that such regulation is required not in order to enable *prediction*, as is the usual goal of the Big Data industries, but to enable *modulation*, that is, the productive individuation of sensibility that 'embraces indeterminacy'.

Simondon also provides tools for thinking about the individuation of software-based media technologies. In much contemporary media studies the development of media has been subsumed into two main theories: convergence and remediation.

Although not here wanting to dispute the claims made by either of these theories, they obviously offer insight into the transition from so-called old analogue media to the new digital media, they fail to offer a convincing account of the dynamic development of media itself. So although it may be the case that some newer media have been *remediated* to resemble television and vice versa, to take one example, the theory of remediation struggles to account for the determining factors required for invention beyond these aesthetic factors, which are limited to the human sensorium. Additionally, in the common theories of convergence (for example, functional, ownership, regulatory, cultural), which are widely taught in media studies, the determination is always from the cultural to the technological. Again, to reiterate, I'm not proposing a strong technological determinism, but I do wish to insist upon the presence of a technological dynamic at play. As I hope I've made clear above, I do also think that the intertwining of this technical dynamic with the regime of the psychosocial is more intimate than ever but not so much that the former is subsumed into the latter.

One text that has a Simondonian spirit (although only coincidentally) is Jonathan Zittrain's *The Future of the Internet—And How to Stop It.* As the title suggests, the focus of the book is regarding a danger, which is that of the closing down of the openness and generativity that were the hallmark of the early Internet and which enabled its speedy growth, both in breadth of application but also scale of use. By openness and generativity Zittrain is referring to the general ease by which hobbyist and non-professional computer users can develop and extend the functionality of computing machines. As Zittrain notes, this kind of generativity was actually made possible by the separation of 'software creation from hardware construction' (Zittrain, 2008: 14). As such it resonates with the idea of informational openness that Simondon stipulates as a requirement for technicity.

Zittrain presents a history in which the generative potential of hardware is unlocked by the separation of software from hardware, and the generative potential of the Internet is enabled by the shift from the proprietary 'walled garden' services such as Compuserve and America Online to the more open Internet. This openness led to an explosion of software development and sharing as well as the invention of a range of technologies that have become widely used. However, Zittrain continues, there is a very real danger of serious restrictions to this openness, in part due to the interference of non-technicians:

> Developments then take a turn for the worse: mainstream success brings in people with no particular talent or tolerance for the nuts and bolts of the technology, and no connection with the open ethos that facilitates the sharing of improvements. (Zittrain, 2008: 150)

One target Zittrain has in mind is the development of apps on mobile devices and tablets whose use is limited to a closed set of functions. Apple is particularly draconian regarding which apps they'll allow on their App Store and apps are also very restrictive regarding the utilization of data from other applications. Additionally, there has been a revival of the concept of delivering software as service following the success of so-called Web 2.0.[8] The danger Zittrain observes is the preponderance of locked-down hardware and software, which severely restricts the ease by which development can proceed.

Zittrain's (2008) argument regarding generative development expresses a similar notion to Simondon's affirmation of the importance of indeterminacy for further invention. For Zittrain a platform is non-generative or hypertelic when it is so locked down that it prevents innovation by its users because they are denied sufficient access to development tools or APIs. Conversely a generative platform allows users to 'tinker' and explore freely and thus uncover potential lines of development on that platform. It's true that one of the reasons for such closure is the dangers that can emerge when a system is generative. Zittrain identifies computer viruses as a key example of a development that has led to consequent restriction. This in turn can have normative consequences:

> Generativity instigates a pattern both within and beyond the technological layers of the information technology ecosystem. (Zittrain, 2008: 64)

The importance of margins of indeterminism is identified by Zittrain as crucial for future developments in software and the Internet. Additionally, new technological developments often lead to new ethical problems and one requirement is to provide answers to these that are satisfactory on a technical level as well as the cultural.

Although necessarily sketchy I hope that I have managed to indicate some of the promising routes for development that Simondon provides. I'm aware that the suggestion may appear both broad and unfocused but this is necessarily a function of the project to axiomatize the human sciences. Such an encyclopaedic scope necessarily entails universal application and transformations.

In the final section of this chapter I will attempt to distil Simondon's work into some broad themes in order to enable a clearer understanding. To be clear, this is not a proposal for a Simondonian method but an indicator for how his work can lead future investigations into media technology. The broad themes I'll outline here are between what we could describe as two kinds of technology: those that stimulate invention and those that encourage regulation.

INVENTION AND REGULATION

When considering Simondon's philosophy of technology we can, broadly speaking, divide his consideration between invention and regulation, a split that reflects the allagmatic ontological recognition of operation and structure. Of course, we must remember that structure here is an outcome of operation, thereby making clear that the starting point for his theory is ontological and takes seriously the role of causation.

It is necessary to discuss a little more regarding politics given that this is historically such an important component of media and communication theory, especially that developed from the tradition of British cultural studies.

As we saw from the description of the development of the human relation to the world from that of magical unity, the modern form of politics developed in conjunction with the development of industrial technology in a second phase-shift.[9] As such this kind of political thought, I would argue, is of interest mainly for how it influences the balance between technicity and culture. It would be incorrect to look for support from Simondon for any of the major competing political theories given that he finds them to be built on poor ontological foundations. I have already described his problems with Marx's theories of alienation and labour as being too hylemorphic, for example. It would also be quite easy to launch a critique of the modern liberal subject for being too atomistic. So although his work is in dialogue with

Marx and others regarding labour and technology, the aim is not to formulate a competing or concurrent political theory.

Simondon's project is not concerned with the application of a certain principle of social organization ('the mythology of a group which is set up as a universalizable doctrine' [Simondon, 2001: 225]) derived from anthropological foundations—whether those be based on the rational individual of the market or the communal imperative of the far left—simply because he does not allow for a human essence to provide the necessary foundations. As we saw with Big Data, to extrapolate a social organization from basic anthropology, whether based on a technical schema or not, is always problematic, as it always misses some aspect of that which overflows this anthropology. Fundamentally, such an imposition will not be able to resist the inventive dynamic of technics that overflows even the broadest human group. As with ethics, the political is a domain that constantly individuates and so to attempt to specify a Simondonian political position would be to tie oneself to one moment in an ongoing individuation.

Although Simondon is interested in societal regulation, it is in order to support what he sees as a fundamental objective, which is technology's ability to move beyond itself through invention and the consequent evolutionary potentials this enables for humanity. If we revisit Simondon's archetypal figure of the technician we will get a clearer idea of the scope of this ambition. The technician's interests are not with the political but with the utilization of potentials in technologies that open up new paths of individuation. Simondon is clear that this process of invention is not performed in accordance with the norms of the society in which the technician works. The technician's concerns are with exploration in the sense of discovering new possibilities and new worlds—Simondon openly admitted his childhood admiration for the work of Jules Verne. This kind of exploration occurs beyond the regulation of modern politics; in fact it is productive of new values and norms, which the explorer returns to the collective thereby changing it.

What is the drive for this invention and exploration? It is wonder, imagination, possibility. There is a purity in Simondon's description of the technician that may also strike one as a naivety. But as I hope I've made clear this is not just invention for invention's sake without regard for consequences. There is a coherent ethico-aesthetic aspect to Simondon's theory, which includes the recognition that inventions should be productive of unities of operation that balance with the environment in a non-destructive manner. This is evidenced in his disapproval of artificiality and machine slavery, that is, his ethics and aesthetics disapproves of any situation that denies an individual access to the full use of their operational abilities.

Simondon is therefore sensitive to the environmental impact of technology and may have sympathy with the current term *sustainable development* as long as it is understood as the ability to sustain further invention and not an argument to try and attain a sustainable equilibrium. The latter would be far too conservative for a thinker who promotes *major* technologies for their evolutionary capacities.

And this is a point worth considering when looking at our contemporary media situation. Zittrain is correct when he points out the dangers of non-generative software for the development of the Internet. Certainly we can appreciate social networks, such as Facebook, for their innovation. Like the operation of the market system described above, it is a genuine concretizing invention that has an operative unity constituted through the capture of affect and the psychosocial. But having said this, it is also not a place to get stuck! It is quite clear that as a platform Facebook suffers from the over-determination of capital through advertising and as such encourages only limited, inter-individual relations that bypass the need for a technical understanding at all. If one of Simondon's aims is to save technology from degrading usage then this subsumption to capital-driven marketing and social hypertelia via Big Data is surely worthy of critical interest. In this example we can both witness the profundity of technology in that its use can transform the nature of the social relation but also how it leads to new forms of capture and closure.

It is no coincidence that control is a watchword for those who wish to understand contemporary society given that it is a key term for the technical schema of cybernetics that is paradigmatic of our time. For this reason it is worth making the distinction between regulation and control, where if the latter promises certainty as long as the data is rich enough, the former recognizes and understands that all such promises are misplaced because every structure operates between determinism and indeterminism, that one cannot control that whose metastability means further individuation is inevitable.

An important part of Simondon's work is an ontology that carefully describes this condition of being that, as we saw, corresponds so well with contemporary work in complexity. The importance of this is further deepened by the way he develops this ontology across the regimes of the physical, vital and transindividual, something that Massumi recognizes in his work on affect. However, just as significant is the role of epistemology in Simondon's work, for any account of ontogenesis must also include the development of thought. Not least is his achievement of linking the operation of technical structures to schemas that guide thought as well as our relation to the world. As such we can understand the sense of how epistemology follows technical invention and how this also entails that thinking must be careful in misapplying such schemas. As I've argued above this is precisely the mistake that Pentland has made in transposing the network schema to the social. The danger is that one will always miss that which overflows the scope of such epistemologies.

What fascinates Simondon is the role technologies have in the development of civilizations, both in their material structures but also in the development of their thinking. It is for this reason that Simondon insisted that a revised humanism needed to be a genetic encyclopaedism. In respect to the human sciences this also requires taking into consideration the ontogenetic development of ethics and aesthetics via an understanding that includes the affective-emotional and cognitive. It is in this respect that we can come to an understanding of a more accurate meaning of regulation, that is, as Metrion or *due measure* that balances the whole range of operations under consideration at all levels of any given structural unity. In regards to human activity this necessarily involves both ethical and aesthetic regulation.

ACCELERATIONISM

I would like to finish by indicating some areas of contemporary interest in relation to which I think Simondon would be a productive interlocutor. It is easy to see in the proposals for a revitalized accelerationist project the engagement with the kind of ambitions for the future that Simondon champions with his theory of 'major technology'. That is, it recognizes that technical gestures, especially those at a scale extending beyond the intra-groupal, 'engages the future' through a feedback action from the transformed environment onto humanity. For Simondon such gestures are the very mechanism by which human evolution occurs, but crucially, an important aspect of this development is the concurrent engagement of culture with technicity. One can see in some of the accelerationist texts the same hope for a redefinition (or even eradication) of labour made possible by technical developments. Like Simondon, accelerationism proposes a reconsideration of humanism, or at least of the human. The call to 'reconnect its roots in the Enlightenment, in a rationalist and universal vision' also resonates with aspects of Simondon's project. He is undoubtedly concerned with reworking the enlightenment project, his proposal for a new ontogenetic encyclopaedism being an explicit reference to that most enlightened of projects. Additionally, without wanting to overextend the comparison, the nature of the rationalism that Simondon proposes is rooted in an allagmatic seizing of ontogenetic operations as evinced by technical mentality. As such there is a clear area of shared ground with Reza Negarestani's project of *inhumanism*, for example, the description of which by Mackay and Avanessian, could be of Simondon's project itself:

> What is specific to the human is its access to the symbolic and sociotechnological means to participate in the construction and revision of norms; the task of exploring what "we" are is therefore an ongoing labour whose iterative loops of concept and action yield 'non-monotonic' outcomes. In this sense, understanding and committing to the human is synonymous with revising and constructing the human. (Avanessian and Mackay, 2014: 30)

Simondon was also critical of capitalism and its related consumerist tendency to overburden technology with the 'venality' of 'psycho-social determination'. He would likely prefer an acceleration of the development of a technical culture that would more ably understand the role of technicity and in doing so reduce alienation, as well as improve the appreciation for technicity and prepare more encouraging conditions of technical invention. Such a cultural shift may well encourage a less consumerist outlook and a transformation of work. That is, it would resist the subsumption of technology by capitalism. Undoubtedly what Simondon is proposing is the central importance of technical invention for a concomitant progression in thought via technical schemas and affective modalities.

I will, however, stop short of describing Simondon as an accelerationist. The term has too many connotations with the idea of breaking capitalism through some kind of catastrophe, which isn't present in Simondon's work. Having said this, the technical dynamism Simondon describes is already one of continual individuation and as such implies that further processes of dephasing and subsequent phase transitions are unavoidable.

In some ways Simondon's theory of technological and culture development is radical in its understanding of the evolutionary potential of technology and the requirement to keep moving forward. On the other hand his writing can appear conservative given that his ultimate answer is one of education and regulation.[10]

NOTES

1. For an excellent discussion of this disagreement see *New Media: A Critical Introduction* (Lister et al., 2009).

2. The choice of Hodkinson's text was completely arbitrary and I'm aware it may seem a little unfair to use a textbook for an example. However, as a textbook it does also clearly demonstrate the culturalist perspective endemic in how contemporary media studies is taught. Another recent example of this bias is found in *Misunderstanding the Internet* (Curran, Fenton and Freedman, 2012) in which the authors counter explanations of the Internet's 'impact' being due to technology by stating that these explanations 'failed to grasp that the internet's influence is filtered through the structures and processes of society. This explains, it is argued, why the influence of the internet has varied in different contexts' (179). In this example society predominantly stands in for political economy, which the authors see as the key domain for transformation.

3. We direct readers to Manual DeLanda's excellent *War in the Age of Intelligent Machines* for an account of the history of war considered as the history of mechanical evolution. As such human intention is arbitrary given that the real historical force is that of the rise of mechanical intelligence.

4. For a longer discussion of Big Data understood using Simondon see my article 'Simondon and Big Data' (Mills, 2015).

5. Although this is not the place to expatiate on the appropriate levels and orders of magnitude for such an account of contemporary media, I will suggest a short litany of potential candidates: physical infrastructure, software infrastructure, devices, apps and software applications, psychosomatic individuals, collectivities, institutions and laws as well as operative unities that can be composed of, modulated and organized from combinations of the above. As

such rather than understanding contemporary media as a convergence it is more appropriate to conceive of it as being involved in a multiplication of relations and unities at multiple levels.

6. This might seem to indicate a problem with Simondon's attempts to keep the mode of technical individuation purified from cultural influences; however, this is mistaken. As we have seen with Simondon's theory of the image-cycle, the human is intimately involved with technical invention. Why then this concern with disallowing cultural influence in the first part of *The Mode of Existence of Technical Objects*? Simply, Simondon's concern in that part of the book was with establishing the mode of technical individuation as a tendency in its own right. He thus needed to demonstrate the possibility of independence of this mode of individuation. Such independence of the individual from that which it is individuated enabled Simondon to escape the anthropological tendency in theorizing technology, which was not compatible with his ontogenetic account.

7. Network speeds, along with processor speeds, are constantly being improved; for example, in 2006 the Philadelphia Stock Exchange relocated most of its trading engines '80 miles—and three milliseconds—from Philadelphia, and into NJ2, where . . . the time to communicate between servers is down to a millionth of a second' (Vanderbilt, 2009). These types of improvements are not concretizations, however, as they do not lead to further qualitative developments of the system.

8. For a more detailed description of Web 2.0 see 'Cultural Anxiety 2.0' (Everitt and Mills, 2009).

9. There is a striking similarity here with Kittler's account of the development and transformation of the subject corresponding with that of media technology, which would reward closer investigation.

10. As an aside, I also think that Simondon would be a useful resource for those interested in xenofeminism. In Simondon's allagmatics we can identify clear alignments with the main thrust of the manifesto. With his axiomatic of individuation Simondon offers a universalism without essences, an immanence within transcendence composed of 'perpetual modification'. He also identifies the continual production of new alienations—albeit those associated with technologies and the subsequent requirement for the 'construction of freedom'. Simondon may not be so quick to say that nothing is sacred, but this would be due to the particular nuance he brings to the word and the problems he has with artificiality.

Chapter Nine

Situating Simondon

In this book, due to the overriding interest in media technology, Simondon's work has been approached mainly via the influence of cybernetics. In this chapter, and with the aim of not doing a disservice to the scope of his project, I will discuss some of Simondon's other influences in order to more broadly situate him within a more diverse philosophical landscape. Indeed, Simondon's work is very much alive to dialogue with a wide range of thinkers.

At the heart of Simondon's work is the notion of the individual and he engages with the philosophical question of the individual in detail in his 'History of the Notion of the Individual', which was published as a complement to his main thesis of 1957. In this text Simondon traces the history of the philosophical idea of the individual in depth, predominantly throughout Greek thought but also in relation to some modern figures of Western philosophy, notably Descartes, Leibniz and Malebranche. This engagement illustrates the depth of Simondon's interest in the philosophical notion of the individual, which underpins his critical assessment of hylemorphism, atomism and Gestalt theory. Indeed, one of the ways Simondon understands his reformed notion of information is as a hybridization of the notion of Aristotelian form taken as modulation along with the reformation of Plato's *idea* as the analogue of causal organization in thought.[1]

Beyond this engagement with the notion of the individual Simondon was clearly also responding to and influenced by many other modern philosophers. In the following I want to indicate some of these areas of influence as a point of reference.

NIETZSCHE

Perhaps unsurprisingly, given their shared interest in a renewed sense of humanism, there are elements of Nietzscheanism in Simondon's philosophy as well as some clear references to him in his work. This influence is most obvious during those passages in which Simondon describes an individual being able to sufficiently disassociate itself from its past in order to individuate further. That is, Simondon finds inspiration in those moments in Nietzsche's work where the actions of an *untimely* individual act as a singularity for the invention of a future, rather than as a mere adaptation to a pathological present. There are two instances in Simondon's work where this influence is particularly evident.

The first and clearest example is Simondon's interpretation of Zarathustra's encounter with the fallen tightrope walker. Simondon retells the story such that it is the tightrope walker's death that enables Zarathustra, despite having been living in solitude, to finally disindividuate from the collective at the level of inter-individuality, thereby enabling him to reindividuate at the higher transindividual level. That is to say that the death of the tightrope walker profoundly transformed Zarathustra's understanding of who and what he was. No longer did Zarathustra just understand him via his merely functional properties *qua* tightrope walker, but his death enabled the revelation of a more profound relation to emerge, which given its commonality consequently also transformed Zarathustra's relation to the collective. This transindividual relation can thus be understood as having been obscured by an inter-individual normativity.[2]

The second example is that of the technician, who like Zarathustra is described as existing outside or on the periphery of the collective, but on returning with his invention initiates a process of collective transformation. In this example the introduction of the technical invention inaugurates a structural change of the collective that goes beyond a mere adaptation to prevailing conditions, a state that Nietzsche understood as pathological in the sense that it was a kind of paralysis conditioned by the past. In his theory of the will Nietzsche is interested in those moments when the spell of the past, in exerting a conditioning normative effect on the present, is broken by something untimely, enabling a succession to a future through an unforeseeable event. For Simondon technical activity is understood in just such an evolutive manner and as inaugurating this kind of normative break.

In this manner the individual can also be understood as a relation between two situations. In another striking example Simondon discusses the lifecycle of Coelenterates (such as coral and Medusa) in which different entities in the colony perform different functions. For Simondon the individual is not specified as any one of these individual organisms nor as the colony as a whole, rather the individual is that organism that leaves one colony and by doing so initiates another:

> The individual is not a part of a colony; it fits between two colonies without being integrated into any, and its birth and its end balance inasmuch as it emanates from a community but engenders another; *it is relation.* (Simondon, 2013: 169)

SPINOZA

It is also unsurprising that throughout his oeuvre Simondon makes scattered references to the great philosopher of monism, even if these are sometimes a little opaque. Simondon's ontology can be understood as a kind of monism without the substance, a term that is too tainted by its association with hylemorphism, although Simondon may be being rather harsh in reading Spinoza in this manner.

Spinoza's description of substance specifies that it has two powers, the power of acting and the power of thinking everything it brings into existence. These correspond to the two sides of individuation in Simondon, the transductive individuation of being and the parallel transductive individuation of thought. By contrasting his account of individuation, that is, the ongoing individuation of pre-individuality, with Spinoza's substance, Simondon is attempting to elaborate the difference between a monism that names as substance that which stands as a superior unity in some sense already given and beyond the world and that which is the operation of individuation immanent to itself. As such, individuation for Simondon is situated between the principles of transcendence and substantial monism in a kind of abeyance, which is the operation of relation.

A proximity to Spinoza is also reflected in the importance affect has for Simondon, which is, of course, central to Spinoza's ethics. There is also a Spinozan flavour to Simondon's account of the overcoming of alienation if read as demanding an affective engagement that increases an individual's power to act. As I will describe in more detail below, Simondon's formulation of individuation holds a striking resemblance to the powers account of causality that is also reminiscent of Spinozism.

FRENCH EPISTEMOLOGY

Both Georges Canguilhem and Maurice Merleau-Ponty were Simondon's doctoral supervisors, a propinquity that places him in the lineage of French epistemology and phenomenology. Although this influence is significant, throughout his work Simondon's engagement with cybernetics as well as his ontological commitment to individuation means his position stands rather obliquely to it. Simondon is significant in this tradition by being the first to make the leap to prioritize ontology.

BERGSON

Despite their many differences the influence of the philosophy of becoming developed by Henri Bergson can be felt throughout Simondon's project. Notably, Simondon shares Bergson's refusal of the reductive thinking of becoming to being, that is, that becoming can be properly thought through fully individuated objects. For both thinkers what is fundamental is an onto-genesis that must occur prior to thinking or intentionality. Thus both share a position opposed to both Kant and Husserl. For Bergson, contra Kant, our knowledge can reach the real but unlike Husserl intentionality is not seen as fundamental.

Bergson navigates a path between mechanism and vitalism through a durational 'spiritualist' philosophy that understands novelty and indeterminism as central properties of becoming. This notion of becoming is used to counter traditional quantitative science, which Bergson argues spatializes that which is properly durational. An exception to this is the biological sciences that Bergson argues expressed 'something that is genuinely posed or "intended" by nature' (Gutting, 2005: 57) as opposed to the quantified sciences that he understood as constructed.

We can thus see how important Bergson is for Simondon in that he also eschews an ontology of substance for one of becoming and also adopts a strictly non-reductive, non-mechanistic position with a strong interest in the biological and the complex. However, where Bergson's is an ontology of 'forces' or 'tendencies' (Gutting, 2005: 56), Simondon's is one of operations.

Despite these similarities Simondon does utilize the physical sciences to move away from the more spiritualistic aspect of Bergson's thought. Where Bergson never really escapes from vitalism due to his retention of the *elan vital*, Simondon makes a clean break by rooting his ontogenetic approach in the thermodynamically inspired pre-individual.

Another significant aspect of Bergsonian thought to which Simondon's work responds is his theory of the image. The image has an important place in French thought as a means of overcoming both Cartesian dualism and the phenomenological prioritization of subjectivity. Simondon's reworking of the role of the image was, to some extent, a response to both Bergson and Sartre, who both placed great importance on it.

Bergson's use of image is ontological in that matter is understood as an accumulation of images. The difference between mind and matter then becomes one of degree, described in terms of dilation and contraction. Simondon also utilizes these terms to distinguish the difference in mode of individuation between his regimes but does so in such a way that the image emerges as an activity within the organism, thus avoiding Bergson's implicit idealism.

Sartre's theory of the image also opposes Bergson's mixing of matter with consciousness. On the contrary, Sartre's use of the image aims to delineate the distinction between these two terms. As such, for Sartre, the image is a certain kind of consciousness and consciousness is distinct from the in-itself. As we've seen, Simondon's monism provides an account of the image that is immanent to the individuation of being. For him the imagination genetically develops within the organism from a 'pre-conscious fund of images' whose existence, like Bergson's notion of the image, is independent from any consciousness. As we will see shortly, Simondon's account is far closer to that of the later work of Merleau-Ponty.

BACHELARD AND CANGUILHEM

Simondon's ongoing dialogue with Kant is an extension of that undertaken by his forebears in the post-Kantian enterprise of French epistemology.

As well as being interested in how modern physics enabled the overcoming of the philosophical notion of substance in favour of that of relation, Gaston Bachelard's influence on Simondon can perhaps be more strongly felt through the concept of phenomenotechniques, which holds that science isn't merely a descriptive enterprise but is actually productive of new phenomena, especially through the implementation of technology. This constructivism is also epistemically productive in that the use of technology in science directly leads to new knowledge and ways of thinking. That is, that the 'categories the mind constructs are relative to this historical situation' (Gutting, 2005: 4). Although this sounds like social constructionism (itself another development we can understand as developing from Kantianism), Bachelard is better understood as having a realist leaning in that his concern is with how scientific knowledge can only ever approximate a complex reality. This obvious Kantianism being that the thing-in-itself can never be fully known; however, this is not due to the necessary structure of the *synthetic a priori* but due to the complex nature of a world whose structures defy precise scientific measurement. As such any deterministic description, such as Newtonian mechanism, must be the result of approximated abstraction.

Bachelard clarifies what he sees as Kant's mistake in hypostatizing Newtonian mechanism (or indeed any other scientific theory) as providing the eternal structure of the categories. However, where Bachelard focuses on scientific epistemology (he even talks of technology as 'theories materialized' [Gutting, 2005: 4]), Simondon's commitment to ontology and individuation permits him to give a more widespread and ambitious description of the interplay of technology with a broader reality. Simondon is also clear about the role technology plays in producing new potentialities as well as its role in epistemology with his notion of technical mentality. The latter is reminiscent of Bachelard's famous notion of epistemic breaks, which influenced Foucault (and subsequently Kittler) and that Simondon can be interpreted as ontologizing with his theory of the cycle of the image.

The historical development of knowledge is also a central concern of Canguilhem's (who supervised Simondon's complementary thesis on technology) although, unlike Simondon, he does not propose an ontology to underpin it. It is likely that Canguilhem's influence was quite wide ranging given that he had corresponding interests in the concept of the vital and biology, the production and operation of norms, epistemological breaks as well as the history of the concept of the milieu, which is a central concept for Simondon. Canguilhem also shared Simondon's interest in maintaining the difference between the vital and mechanical, observing that vital organisms have the capacity to create new norms as an active response to their environment, an ability mere mechanisms lack.

MERLEAU-PONTY

Arguably the key figure within the tradition of French epistemology for Simondon was his doctoral supervisor Maurice Merleau-Ponty. As de Beistegui (2005) notes, although there is a scarcity of evidence that there was any dispute between the two philosophers, it is undoubtedly the case that they shared a 'common ambition' insofar as the later work of Merleau-Ponty sought to move away from its Husserlian foundation towards an ontological one. It is worth noting that although Simondon is sometimes described as a phenomenologist, the work of Husserl held little influence on him as he saw the foundational place it gave intentionality as misplaced given that he understood it as arriving quite late on the scene ontogenetically.

It is worth digging a little deeper into Merleau-Ponty's late philosophy to understand the connection between the two thinkers and the radical nature of Simondon's break with phenomenology. In the notes from Merleau-Ponty's course on nature one can discern the groundwork for his later philosophy, which was to be elucidated in the unfinished work *The Visible and the Invisible*, and which traverses some of the same key issues that are found in Simondon's work.

As several commentators have elucidated (Hansen, 2005; Mazis, 2000) Merleau-Ponty's second turn to biology was in response to what he saw as the failure of his first engagement (utilizing Gestalt theory) in that it didn't manage to overcome its foundation on the dualism of consciousness and object. In a working note in *The Visible and the Invisible* he writes:

> The problems posed in Ph.P. [Phenomenology of Perception] are insoluble because I start there from the "consciousness"-"object" distinction. (Merleau-Ponty, 1968: 200)

With the attempt to bridge this gap, that is, to answer the question 'what is the alleged objective conditioning?' (Merleau-Ponty, 1968: 200), Merleau-Ponty moves away from phenomenology and towards ontology. The concept of nature, then, signifies that which is ontologically primordial to the object-consciousness dualism.

At the heart of Merleau-Ponty's exploration is the theorization of embodiment as an organization from which consciousness emerges as a behaviour. That is, consciousness is immanent to the activity of nature (the biological body) and through understanding its emergence we can understand what nature is. As Hansen argues,

> The fundamental correlation of behaviour and morphogenesis Merleau-Ponty discovers in his exploration of the biological sciences grounds the correlation of phenomenology and ontology in his late work, and that it does so precisely because it overcomes the dichotomy between mind and body on one side and world and environment on the other. (Hansen, 2005: 233)

As such Merleau-Ponty isn't as ambitious as Simondon in making an ontological leap beyond the organism. In his earlier work, such as the *Phenomenology of Perception*, Merleau-Ponty predominantly focuses on the perception of the world undertaken by the subject. In his later work he shifts to attempting to describe an ontology that could underpin these earlier phenomenological investigations but that also demonstrates how the gap between consciousness and object is not foundational. Furthermore, one of the tasks he undertakes is to show how this gap is constructed and is a necessary aspect of the morphogenetic development and ongoing existence of the organism.

Simondon also has the erasure of this gap in mind with his assertion of the priority of the topological-chronological dimension for the organism. Simondon's concern is for the overcoming of the dualism such a gap implies, which is unacceptable for his genetic informational ontology based on individuation beginning from the pre-individual. As such he must demonstrate the establishment of the psychic from the regimes of individuation.

A key part of the second course on nature in relation to the development of Merleau-Ponty's later 'ontological turn' is regarding the work of the ethologist Jacob von Uexküll. One concept of von Uexküll's that Merleau-Ponty returns to frequently throughout his later work is that of the Umwelt. Merleau-Ponty describes the Umwelt as that which

> marks the difference between the world such as it exists in itself, and the world as the world of a living being. It is an intermediary reality between the world such as it exists for an absolute observer and a purely subjective domain. (Merleau-Ponty, 2003: 167)

As a concept that deals directly with how the 'consciousness-object distinction' is established it is clear why this concept engages Merleau-Ponty. Von Uexküll considered himself as a Kantian because he maintained that there could be no such thing as a meaningful objective reality because meaning is something that is constructed *between* an organism and its environment; it is a function of its *Umwelt* (we have seen Simondon says something similar when he describes meaning as a relation).

It is the idea of the emergence of consciousness through such a relationship that Merleau-Ponty investigates through the concept of behaviour. The understanding of behaviour is key for understanding how von Uexküll's theory of the *Umwelt* helps Merleau-Ponty ontologically found the separation of the objective and subjective.

Significantly, behaviour is not to be understood as something that occurs only at the level of consciousness but neither is it to be understood as mechanistic. In fact,

> consciousness is only one of the varied forms of behaviour; it must not be defined from within, from its own point of view, but such as we grasp it across the bodies of others; not as a centrifugal form, but as a closed world where external stimulations appear to it as outside of it. (Merleau-Ponty, 2003: 167)

Merleau-Ponty argues that spatio-temporal experience is constructed by the organism's developing relation to its *Umwelt*, that is, it is constructed and contingent upon the organism's perceptual capabilities and behaviour. As such Merleau-Ponty, like von Uexküll, would have no problem asserting that different animals may have radically different experiences of the world. So although von Uexküll would agree with Kant that we couldn't have direct experience of things-in-themselves a certain ontological necessity requires that there actually is a brute world against which each organism constructs its *Umwelt*. The existence of the *Umwelt* therefore indicates the existence of this brute world and forecloses the possibility of absolute idealism.

Likewise, the positing of an *Umwelt* must not be confused with a form of naive realism in that there is no direct perception of a world. The *Umwelt* is developed by the relation of the activity of the organism with the world and as such the *Umwelt* is meaningful to that organism. The *Umwelt* therefore relies on the structure of the organism and also plays a role in guiding the future behaviour of the organism as far as it acts as an individual.

There is clearly a resonance of Simondon's theory of the associated milieu with that of the *Umwelt* along with the notion of the field, which also interested Merleau-Ponty in providing an immanentist explanation for an organisation's behaviour:

> Only a field has properties such that it is always distinguished from things partes extra partes, because it always includes a relation between the parts and the whole. It is a regulative principle. (Merleau-Ponty, 2003: 150)

This field, behaviour as form, is theorized then as something that emerges from the 'lower level' system of criss-crossing motor powers of the body in an epigenetic movement. Merleau-Ponty is careful here not to fall into hylemorphism. The behaviour of an organism must not be thought of as in some way the fulfillment of what is already contained in some kind of essence (entelechy) or vital force that has always been present in the organism as this would be to 'double observed reality with a second reality' (Merleau-Ponty, 2003: 152), which would also require explanation. Instead the organism's future must not be seen as being 'contained in the present' but coming 'from the present itself'.

But how does the future emerge from the present? What is the driver of development? Merleau-Ponty's answer to this is that there is a principle of negativity or absence operating within the organism, which enables development through the creation of an imbalance or disequilibrium ('The negative principle is less identity-with-self than non-difference with self' [Merleau-Ponty, 2003: 156]). This disequilibrium cannot be stabilized by a return to a previous state due to the presence of structures within the organism that disallow it, therefore the disequilibrium must be resolved by development along another path. This form of development is to be understood not as just operating via physiochemical processes but partakes of the notion of the totality of the organism, as theorized above as a field, and elsewhere as a 'dimension'.

It is through the establishment of this totality, the adhesion of the 'elements of the multiple' as well as the negativity inherent within the organism that Merleau-Ponty utilizes the work of these biologists to theorize an emergence that springs immanently from the organism. The development of an organism progresses as a cascade where, as each structure is developed, fresh possibilities for differentiation emerge. This development is ongoing and established not in advance but from within the operation of the process itself. The actual organism is only one possible realization of the potentials that were carried in each part of the organism at prior stages of its development.

With his focus on individuation Simondon is concerned with the development of the individual from the outset. Where Merleau-Ponty moves towards ontology as a means to get beyond the subject-object dualism Simondon takes this as his starting point. For Simondon this dualism, as such, is not the main issue to be resolved. For him all dualisms need to be interrogated in order to provide an ontogenetic account of how they developed from that which is pre-individual.

Thus although there are many similarities between his and Merleau-Ponty's later project, such as the central importance that the ideas of the milieu and field have, their crucial point of difference is the importance of the axiomatic nature of ontogenesis for Simondon. Where Merleau-Ponty describes an organism's ontogenesis as the production of a disequilibrium generated via negativity, Simondon founds his entire philosophy on the notion of *metastability* and the transductive mode of individuation that this enables. As such Simondon's work on psychic and collective individuation and of the cycle of the image can be understood as a reinterpretation of much of that which interested Merleau-Ponty in light of this axiomatic of the problem.

REALISM

The account of the idealism inherent in Simondon's realism, the importance of which Simondon undoubtedly inherits from his grounding in French epistemology, should undoubtedly be of interest for those interested in the recent speculative turn towards realism. Indeed one of the criticisms of this disparate movement is that although it launches a coherent argument for realism against post-Kantian philosophies of correlation, to use Meillassoux's term, it 'seems to have *an impoverished if not naïve grasp of relationships*, particularly the relationship between thinking and being' (Galloway, 2015).

Although I have no desire to defend speculative realism, nor indeed to identify Simondon with that particular movement, if one could call it that, it is the case that Simondon's work provides a coherent realism that directly responds to Galloway's concerns and is thus of contemporary interest. Galloway divides those thinkers he sees as having impoverished grasp of relations into two kinds:

> those who prioritize and fetishize relation to such a degree that it loses much of its meaning (in particular, followers of Deleuze, Whitehead, or Latour), and those who marginalize relation in favor of some other overarching concept (contingency for Meillassoux, or the One for Laruelle). (Galloway 2015)

Clearly Simondon's ontology can be understood as relational in that from the very first dephasing of the pre-individual being it is productive of relation. Does this mean he falls into either of the problematic camps that Galloway stipulates, notwithstanding the validity of his critique?

As usual Simondon's ecumenical approach enables him to encompass both positions while, I believe, not losing its meaning. I think Galloway is correct with his assessment of Latour (I make a similar critique in chapter 7), but although Simondon prioritizes relations it is by way of being clear about what they are and how these comprise the entities that are related. That is to say that the reality of relation does not entail that the entities that are related are any less real.

To achieve this Simondon pays careful consideration the requirement, if a coherent realism is to be built, that the opposite terms of relations must be coherently resolved into one another without losing their sense. This is the genius of Simondon's work that he maintains dualisms while ultimately going beyond them by treating the extreme terms of each dualism as the extremities of a relation from which those terms emerged. His work is unrelenting in describing individuation in all domains as the creation of terms via relation through which the construction of dualisms can be understood by revealing the obscure zone of the work of relation that resides at their centre.

Does he then marginalize relation in favour of some other overarching concept? Certainly the pre-individual would be a candidate but I'd argue that relation isn't marginalized but rather is a condition of the metastability of the pre-individual. As such, relation is an immanence, it is a condition *of* the pre-individual rather than conditioned *by* it. Given such a strong and coherent account of relation I contend that Simondon's ontology evades that depreciation into materialism that Galloway claims would be its fate.

What's more, to respond to Galloway's concerns regarding any inadequacy of the relation between thinking and being it should be quite clear that Simondon provides a strong and intimate account of the biological genesis of the image (in concert with the collective) in *Imagination et invention* as well as a clear method of how thought is related to being in the allagmatic epistemological method.

To clarify my argument further that Simondon's is a viable realism that provides a coherent sense of relation I want to demonstrate this coherence by contrasting his ontogenetic account of individuation with a recent analytic account of causality that argues that relation is in fact produced as part of the causal operation.

CAUSALITY

Causality still remains a vibrant metaphysical topic for contemporary analytical philosophy and has recently undergone a turn to realism in the form of a revival of dispositional or powers-based theories. In this section I want to support my claims that Simondon's ontology is productive of a coherent realism, as well substantiate the various references I've made regarding the similarities that Simondon's ontology has to those who hold a powers-based metaphysics (such as described by Spinoza for example) by favourably contrasting it with some recent analytical contemporary work in this area.

Specifically, I will contrast it to Mumford and Anjum's excellent text *Getting Causes from Powers* (2011) in order to show how Simondon's work provides a logically coherent account of causality as well as an avenue by which this schema can be coherently extended.

This contrast is also useful in that it makes clear the difference between the transductive method and the purely analytical approach, particularly in that although the latter offers an important measure of consistency, it fails to provide the kind of tools required to make the kind of constructive claims of which the former is capable.

Disposition- or powers-based (for the purpose of this discussion these terms will be synonymous) metaphysics is realist and holds, following the Eleatic stranger's test of reality in Plato's *The Sophist*, that something is real if

> it has any capacity at all, either by nature to do something to something else or to have even the smallest thing done to it by even the most trivial thing, even if it only happens once. (247e)

Powers-based metaphysics thus holds that the definition of the reality of an entity is that it has the capacity to do something or have something done to it. An entity's ability to act is called its powers or dispositions, which are also often referred to as properties of the entity. It is impossible, according to this theory, that an entity has no power or disposition (actual or in potentia) for it is its dispositionality that defines its reality as an entity.

Mumford and Anjum defend a dispositional theory called dispositional monism that, as the name suggests, is a theory that causal explanation is given solely in terms of dispositions. It is thereby a true powers-based theory of causation in which dispositions are the sole components of causal explanation. As such the dispositional monist argues that dispositionality is necessarily a *sui generis* modality that Mumford and Anjum admit is a theory of causation but can't be a reductive analysis of causation as powers are already causal notions themselves (Mumford and Anjum, 2011: 7–8).

In this section I want to briefly outline the core aspects of this *sui generis* modality while also aligning it with Simondon's account. Given the importance of causality for explaining events the importance of such an undertaking should be clear. Not least, what is at stake is a challenging of the residual humanism and Kantianism that still haunts the social sciences.

The main characteristics of Mumford and Anjum's description of causality are that it lies between necessity and contingency, that cause and effect occur simultaneously, that it allows for emergence and that it provides an account of grounding. Significantly, this primitive modality does not attempt to prove that causal connection is necessary. Mumford and Anjum believe it is due to a mistaken response to the Humean theory of constant conjunction that philosophers continually attempt to prove the necessity of causal connection (Mumford and Anjum, 2011: 49). Instead they develop a dispositional account of causality that has possibility and necessity as its extreme cases:

> Dispositionality is not necessity and it is not pure possibility but something in between. (Mumford and Anjum, 2011: 183)

The core reason they give against providing a purely necessary theory of causation is that 'a natural process can be interfered with and thus that a cause never necessitates its effect' (Mumford and Anjum, 2011: 12). As such they assert that pure determinism and pure indeterminism are limit cases on a causal spectrum. This corresponds with Simondon's account of information that he describes as 'midway between pure randomness and absolute regularity' (Simondon, 2001: 137).

Simondon's description is made as a consequence of his ontological claim regarding the metastability of the pre-individual. As such Simondon's argument relies on the truth of his claim regarding the metastable nature of being, while Mumford and Anjum's argument is sustained *a posteriori* by the lack of empirical proof of any form of causation that is not preventable.

As such we can see that both theories come to the same conclusion from different premises; Simondon reaches it as a necessary outcome of his ontological claim regarding primary reality and Mumford and Anjum from empirical observation regarding the preventability of causes.

SIMULTANEITY

Mumford and Anjum's causal theory also resonates with Simondon's ontogenetic approach in that both understand causation as a process in which cause is simultaneous to effect.

> We see the causation as an unfolding process whereby *a* turns into *b*. The combined powers of the cause, such as the solute and the solvent, become the effect—a substance held in liquid suspension—as part of what it is to be those powers. (Mumford and Anjum, 2011: 119)

Unlike the Humean understanding of causation as an unexplained relation between two distinct events (cause/effect) this theory understands causation as a single temporal process in which the cause merges into its effect. As Mumford and Anjum (2011: 119) explain, 'If we accept that causation ought to be some kind of relation holding between temporally distinct events then the game already may be lost'. For as soon as the notion is in place that cause and effect are distinct events the problem becomes how to connect these two events in a relation and explain the nature of that relation.

The simultaneity of cause and effect is also central to Simondon's transductive theory of individuation in which causation is described as a temporal process involving the resolution of disparities in a manner that also creates a corresponding chronological and topological dimension relative to that causal occurrence. What's more, Mumford and Anjum here provide an excellent argument for the need to understand relation not as something that needs to be explained by connecting two events but rather that which is created in the process between them. As such Simondon would concur with the powers theorists that

> causation as a coming together of many powers, having an effect as a joint combined manifestation according to some function of composition. . . . The cause will be depicted as merging into and becoming the effect through a natural process.

The key difference, once again, is in how Simondon describes how powers come together in his ontology in terms of disparity and information. However, such an account helps clarify the status of relation as product of an ongoing causal process, thus countering Galloway's assertion that realist approaches, such as Simondon's, necessarily rely on a meaningless concept of relation as *explanans.*

EMERGENCE

In giving an account of non-linear causation Mumford and Anjum suggest an account of causality called *compositional pluralism*, which holds that there is 'a plurality of ways in which powers compose to produce an effect' (Mumford and Anjum, 2011: 86). The underlying idea here is that often dispositions do not work together in a linear and additive manner towards an expected outcome but can combine in non-linear and non-additive ways and produce outcomes that are described as strongly or ontologically *emergent*.

Mumford and Anjum explain emergence as a consequence of compositional pluralism but do also hold to a doctrine of supervenience (the same composition of powers always lead to the same emergent effect). They also subscribe to a theory of levels in which levels are also described as manifestations. In short, from the perspective of dispositional monism and compositional pluralism it is simply the case that when multiple dispositions work together the resulting effects can be strongly emergent, that is, one can get new powers from old. Thus dispositional monism does not attempt an analysis of how emergence operates as a process but positions itself as just being able to accommodate it via compositional pluralism.

We've already described how Simondon's ontology supports emergence. By specifying the metastable nature of the pre-individual as ground he sharpens his explanation of the operation of emergence through phase-shifts. Once again Simondon's account fits with the analytic powers approach; however, once more the specificity of his ontology means that he cannot remain as non-committal as Mumford and Anjum regarding certain consequences. For example, Mumford and Anjum claim that 'it seems to be an asset of the dispositional theory, therefore, that it leaves the question of emergentism versus reductionism open' (Mumford and Anjum, 2011: 103). Simondon certainly does not hold that the question of reductionism is left open as his clear commitment to strong emergence via phase-shifts makes this impossible.

Should we then see Simondon's description of individuation (which is also an account of causation) as a compositional pluralism? Although we can certainly interpret his description of the resolution of disparities as such we should not lose sight that his account concerns an axiomatic of how being operates that is compatible with powers theorizing such as Mumford and Anjum's but is also attempting to do more work in explaining how actual compositional operations work.

POTENTIAL

Another characteristic of dispositional monism is that it must be possible for a disposition to be real even when it's not actualized. That is, it has to be the case that dispositions are 'unactualized possibilia' (Lie, 2009: 128). This requires that a clear distinction is made between a disposition and its manifestation, with the former being in the seemingly curious position of being real without being actual.

> What is the nature and extent of their existence, actuality, or their being when they are unmanifested? This can be called the question of Being, making use of an ancient term. (Mumford, 2006: 481)

If one were to hold the contrasting Megarian position, which holds 'that only manifested properties are real' (Lie, 2009: 121) and thus there is no such thing as 'unactualised possibilia' then one runs into problems in explaining how one can account for certain dispositions when they're not manifest. It is a problem then for the dispositional monist to explain what grounds a disposition, particularly an unactualized one.

Simondon's notion of the pre-individual, which is rich in potential, as well as that any individual is always overflowing with potential, beyond being fully resolved, particularly through relations with a milieu, means that possibility is not a problem to be explained. The reality of potential is a given at the basis of his ontology and is something that is always carried forward in every individuation as a remainder enabling further individuation. This must be the case otherwise we would be in danger of living in a static universe.

Mumford and Anjum can't offer this kind of operational account of how being both changes and harbours potential. We don't offer this as a criticism as both Mumford and Anjum admit that they are not attempting an analysis of causation:

> Causal dispositionalism is a theory, therefore, but not an analysis of causation. We can say more. Powers, we maintain, are productive of their manifestations, and production is clearly itself a causal notion. We cannot, therefore, analyse causation in terms of an already causal notion of production. This we accept. (Mumford and Anjum, 2011: 8)

As a theory it has many attractive features that help explain the parameters within which a theory of causation must work. However, the lack of focus on the operational aspect of the causal process means that this theory lacks the necessary scope to offer a productive schema for understanding certain phenomena, such as, for example, the mode of being of technical objects.

Although Mumford and Anjum hold that powers are irreducible we claim that Simondon points to something more fundamental, which is how powers *operate*. The powers theorist could argue that this isn't a problem for their claim that powers are *sui generis* and irreducible, for they could counter that the more fine-grained account of operation that Simondon offers could also be described using dispositions. I'm unconvinced that such an argument would work. For me, as for Simondon, all powers involve operation and it is operation that is fundamental. For any description at any level of being one could assert that an entity has a certain power but would it not always be then possible to ask *how* that power operates when enacted? In any description an operational description must be possible to describe the actualization of a power. That is, to make the strong claim that without operation, powers are not powers, but schemata of possible powers.

What I believe this comparison between powers theory and allagmatics shows is that it is possible that powers are not the irreducible mode of causality Mumford and Anjum claim if one is prepared to follow Simondon in his axiomatic account of operation. If one does this then one can see that operation (in Simondon's sense) gives a finer-grained account of the working of powers. To do this does require an ontological commitment but also enables powers to be reduced further. This commitment and reduction I argue gives Simondon's ontology real analytic rather than just theoretical power.

From this perspective it is not that powers theory is incorrect or inconsistent—it is a very well argued and logically consistent schema of the mode of causality—but it isn't detailed enough to explain phenomena. To explain causation as the passing around of powers may be a good way to counter Humeanism regarding causation, as well as develop a clear account of the modality of causation, but does not go far enough to establish an ontogenetic account of powers, without which powers metaphysics remains merely formal. However, Simondon enables a more fine-grained account of causation via his allagmatic description of operation.

Both Simondon's and powers metaphysics are theories that are concerned with the modalities of being. However, due to its development from concepts such as metastability and phase-shifting Simondon's understanding of being engenders a different approach to the modality of causation to that of powers theory. Although Simondon doesn't claim to be a dispositionalist, there is no doubt that his concern is with the power being has *to do*. As I've explained, the heart of his philosophy concerns the *operation* of being. To illustrate this I can point to the different emphasis each theory puts on the notion of form. Simondon's is primarily a theory of the development and transformation of form. This concentration on the genesis of form is closely linked with Simondon's opposition to the hylemorphic schema of explanation. It is through his reformulation of how form arises that his understanding of causation and relation are acquired. That is to say that Simondon's understanding of cause and relation follow from the transductive account of the ontogenetic development of form.

Although powers theory as described by Mumford and Anjum is concerned with how things change it is actually non-committal regarding the mode of operation of ontogenetic development. Their main concern is with countering the Humean twin-event theory of causation rather than giving an account of ontogenesis. The closest Mumford and Anjum come to discussing the development of form is through the theory of the passing around of powers, which has little to say about the role of modulation or organization, for instance.

Simondon, however, makes a firm commitment regarding the operative nature of being in ontogenesis and such a commitment has consequences. This commitment arguably puts him at similar risk to that which we witnessed with Kant's commitment to mechanism, that is, on committing to the truth of a specific scientific theory. In this case Simondon founds his notion of pre-individuality on the theory of metastability adapted from thermodynamics and quantum mechanics. Simondon's theory of becoming is certainly consistent with much contemporary scientific thinking regarding complexity and the individuation of physical phenomena. But that is not to say that such scientific theories won't develop and change over time.

However, the core Simondonian insight of the presence of disparity inherent in metastable being, which acts as a motor of ontogenesis, is a crucial one and although inspired by scientific theories is not dependent on them. That is to say that the claim made for the operationality of being is resistant to the development of such theories for it points to the metaphysical reality of operation irrespective to how it is described by science.

What Simondon's commitment gives him is a richer operational account of ontogenesis than that given in the powers theories we have looked at. What it offers is not thereby inconsistent with powers theory, but where powers theory remains merely formal and in some cases non-committal, Simondon's allagmatics commits itself to being 'in the order of the sciences' (Simondon, 2013: 529) and is thus able to have real explanatory power due to its axiomatic nature.

NOTES

1. The Plato we have in mind here is that of one-world Platonism involving 'idealism's realism concerning Ideas' (Dunham, Grant and Watson, 2011: 6). This is a realism that conceives of the idea as an invariable law, which both attempts an analogy of physical operation and is itself *qua* idea a part of the causal universe.

2. Deleuze elaborates in a similar way on the Dickensian tale from *Our Mutual Friend* about the death of a rogue that causes in the observing crowd a collective empathy and 'who attains a sort of beatitude' (Deleuze, 2001: 29).

Bibliography

All quotations from Simondon 1964, 1989, 1995, 2001, 2008, 2013, and 2014 have been translated by the author.

Achterhuis, H. (Ed.). 2001. *American Philosophy of Technology: The Empirical Turn*. Bloomington: Indiana University Press.

Anderson, C. 2008. "The End of Theory: The Data Deluge Makes the Scientific Method Obsolete." *Wired*. Available at http://www.wired.com/science/discoveries/magazine/16-07/pb_theory.

Aristotle. 1991. *The Complete Works of Aristotle*. Edited by J. Barnes. Princeton, NJ: Princeton University Press.

Ashby, R. 1954. *Design for a Brain*. 1st ed. New York: John Wiley and Sons, Inc.

Ashby, R. 1957. *An Introduction to Cybernetics*. 2nd ed. London: Chapman and Hall Ltd.

Avanessian, A., and R. Mackay (Eds.). 2014. *#Accelerate: The Accelerationist Reader*. Falmouth: Urbanomic.

Ball, P. 2009a. *Shapes: Nature's Patterns: A Tapestry in Three Parts*. 1st ed. Oxford: Oxford University Press.

Ball, P. 2009b. *Flow: Nature's Patterns: A Tapestry in Three Parts*. 1st ed. Oxford: Oxford University Press.

Ball, P. 2009c. *Branches: Nature's Patterns: A Tapestry in Three Parts*. 1st ed. Oxford: Oxford University Press.

Barthélémy, J. 2010. "What New Humanism Today?" *Cultural Politics: An International Journal*, vol. 6, no. 2, pp. 237–52.

Bateson, G. 2000. *Steps to an Ecology of Mind*. Chicago: University of Chicago Press.

Bigelow, J., A. Rosenblueth, and N. Wiener. 1943. "Behavior, Purpose and Teleology." *Philosophy of Science*, vol. 10, no. 1, pp. 18–24.

Bruegger, U., and K. Knorr Cetina. 2002a. "Inhabiting Technology: The Global Lifeform of Financial Markets." *Current Sociology*, vol. 50, no. 3, p. 389.

Bruegger, U., and K. Knorr Cetina. 2002b. "Traders' Engagement with Markets: A Postsocial Relationship." *Theory, Culture and Society*, vol. 19, no. 5/6, p. 161.

Chabot, P. 2013. *Simondon's Philosophy: Between Technology and Individuation*. New York: Bloomsbury.

Combes, M. 2013. *Gilbert Simondon and the Philosophy of the Transindividual*. Cambridge, MA: MIT Press.

Cordella, A., and M. Shaikh. 2006. "From Epistemology to Ontology: Challenging the Constructed 'Truth' of ANT." *LSE Working Paper Series,* Department of Information Systems, LSE, London.

Curran, J., N. Fenton, and D. Freedman. 2012. *Misunderstanding the Internet.* London and New York: Routledge.

de Beistegui, M. 2005. "Science and Ontology: From Merleau-Ponty's 'Reduction' to Simondon's 'Transduction.'" *Angelaki,* vol. 10, no. 2, pp. 109–13.

DeLanda, M. 1991. *War in the Age of Intelligent Machines.* New York: Zone Books.

DeLanda, M. 2011. *Philosophy and Simulation: The Emergence of Synthetic Reason.* London: Continuum Books.

Deleuze, G. 2001. *Pure Immanence: Essays on a Life.* New York: Zone Books.

Deleuze, G. 2004. *Desert Islands and Other Texts 1953–1974.* Los Angeles: Semiotext(e).

Dumouchel, P. 1995. "Gilbert Simondon's Plea for a Philosophy of Technology." In *Technology and the Politics of Knowledge,* edited by A. Feenberg and A. Hannay, pp. 255–71. Evanston: Indiana University Press.

Dunham, J., I. H. Grant, and S. Watson. 2011. *Idealism: The History of a Philosophy.* Durham: Acumen.

Dupuy, J-P. 2009. *On the Origins of Cognitive Science: The Mechanization of Mind.* 2nd ed. Cambridge, MA: MIT Press.

Ellul, J. 1964. *The Technological Society.* New York: Vintage Books.

Everitt, D., and S. Mills. 2009. "Cultural Anxiety 2.0." *Media, Culture and Society,* vol. 31, pp. 749–68.

Feenberg, A. 1996. *From Essentialism to Constructivism: Philosophy of Technology at the Crossroads.* Available at http://www.sfu.ca/~andrewf/talk4.html.

Feenberg, A. 1999. *Questioning Technology.* New York: Routledge.

Feenberg, A. 2002. *Transforming Technology: A Critical Theory Revisited.* New York: Oxford University Press.

Ferré, F. 1995. *Philosophy of Technology.* Athens: University of Georgia Press.

Fuller, M. 2005. *Media Ecologies: Materialist Energies in Art and Technoculture.* Cambridge, MA: MIT Press.

Fuller, M. (Ed.). 2008. *Software Studies: A Lexicon.* Cambridge, MA: MIT Press.

Galloway, A. 2004. *Protocol: How Control Exists after Decentralization.* Cambridge, MA: MIT Press.

Galloway, A. 2015. "Assessing the Legacy of That Thing That Happened After Poststructuralism." *Culture and Communication.* Available at http://cultureandcommunication.org/galloway/assessing-the-legacy-of-that-thing-that-happened-after-poststructuralism.

Grosz, Elizabeth. 2012. "Identity and Individuation: Some Feminist Reflections." In *Gilbert Simondon: Being and Technology,* edited by A. De Boever et al., pp. 37–56. Edinburgh: Edinburgh University Press.

Gutting, G. 2005. *Continental Philosophy of Science.* Oxford: Blackwell Publishing.

Hansen, M. B. N. 2005. "The Embryology of the (In)visible." In *The Cambridge Companion to Merleau-Ponty,* edited by T. Carman and M. B. N. Hansen, pp. 231–34. Cambridge: Cambridge University Press.

Hansen, M. B. N. 2006. "Media Theory." *Theory, Culture and Society,* vol. 23, no. 2–3, pp. 297–306.

Hansen, M. B. N. 2009. "System-Environment Hybrids." In *Emergence and Embodiment: New Essays on Second-Order Systems Theory,* edited by B. Clarke and M. B. N. Hansen, pp. 113–42. Durham: Duke University Press.

Hansen, M. B. N. 2011. "From Fixed to Fluid: Material-Mental Images Between Neural Synchronization and Computational Mediation." In *Releasing the Image: From Literature to New Media,* edited by J. Khalip and R. Mitchell, pp. 83–111. Stanford, CA: Stanford University Press.

Hansen, M. B. N. 2012. "Engineering Pre-Individual Potentiality: Technics, Transindividuation, and 21st-Century Media." *SubStance,* vol. 41, no. 3, pp. 32–59.

Hansen, M. B. N. 2015. *Feed-Forward: On the Future of Twenty-First-Century Media.* Chicago: University of Chicago Press.

Hart, J. 1980. "Preface." In *On the Mode of Existence of Technical Objects,* translated by Ninian Mellamphy, pp. i–xxiii. London, ON: University of Western Ontario.

Hayles, N. K. 2002. *Writing Machines.* Cambridge, MA: MIT Press.

Heidegger, M. 1966. *Discourse on Thinking.* New York: Harper & Row.

Heidegger, M. 1977. *The Question Concerning Technology and Other Essays.* 1st ed. New York: Garland Publishing, Inc.

Hodkinson, P. 2011. *Media, Culture and Society*. London: Sage.

Kant, I. 1987. *Critique of Judgment.* Translated by Werner S. Pluhar. Indianapolis: Hackett.

Kant, I. 1999. *Critique of Pure Reason.* Translated and edited by Paul Guyer and Allen W. Wood. Cambridge: Cambridge University Press.

Kauffman, S. 2010. *Reinventing the Sacred: A New View of Science, Reason, and Religion.* New York: Basic Books.

Kittler, F. 1997. *Literature, Media, Information Systems: Essays.* New York: Routledge.

Kittler, F. 1999. *Gramophone, Film, Typewriter*. Stanford, CA: Stanford University Press.

Kramer, S. 2006. "The Cultural Techniques of Time Axis Manipulation: On Friedrich Kittler's Conception of Media." *Theory, Culture and Society,* vol. 23, no. 7–8, pp. 93–109.

Larrain, J. 1979. *The Concept of Ideology*. London: Hutchinson.

Latour, B. 1993. *We Have Never Been Modern.* Translated by Catherine Porter. Cambridge, MA: Harvard University Press.

Latour, B. 2002. "Morality and Technology: The End of the Means." Translated by Couze Venn, *Theory, Culture and Society,* vol. 19, no. 5, pp. 247–60.

Latour, B. 2012. "Biography of an Investigation: On a Book about Modes of Existence." Translated by Catherine Porter, *Social Studies of Science*. Available at http://www.bruno-latour.fr/sites/default/files/126-KARSENTI-AIME-BIO-GB.pdf.

Latour, B., and M. Callon. 1981. "Unscrewing the Big Leviathan: How Actors Macro-Structure Reality and How Sociologists Help Them Do So." In *Advances in Social Theory and Methodology: Toward an Integration of Micro- and Macro-Sociology*, edited by K. Knorr Cetina and A. V. Cicourel, pp. 277–303. Boston: Routledge and Kegan Paul.

Lie, S. A. N. 2009. *Naturalness Reconsidered: Ontology, Ethics and Dispositions*. Tromsø: University of Tromsø.

Lister, et al. 2009. *New Media: A Critical Introduction*. Oxon: Routledge.

Mackenzie, A. 2006. *Cutting Code: Software and Sociality.* New York: Peter Lang Publishing.

Marcuse, H. 1991. *One-Dimensional Man: Studies in Ideology of Advanced Industrial Society.* 2nd ed. London: Routledge.

Marcuse, H. 2001. "The Containment of Social Change in Industrial Society." In *Towards A Critical Theory of Society,* edited by Douglas Kellner, pp. 83–94. London: Routledge.

Marx, K. 1988. *Economic and Philosophic Manuscripts of 1844.* New York: Prometheus Books.

Massumi, B. 2015. *The Power at the End of the Economy.* Durham: Duke University Press.

Massumi, B., A. De Boever, A. Murray, and J. Roffe. 2010. "'Technical Mentality' Revisited: Brian Massumi on Gilbert Simondon." *Parrhesia,* vol. 1, no. 7, p. 36.

Mazis, G. A. 2000. "Merleau-Ponty's Concept of Nature: Passage, the Oneiric, and Interanimality." *Chiasmi International,* vol. 2, no. 2, pp. 223–24.

McCullough, W. S., and W. H. Pitts. 1943. "A Logical Calculus of the Ideas Immanent in Nervous Activity." *Bulletin of Mathematical Biophysics,* vol. 5.

Merleau-Ponty, M. 1968. *The Visible and the Invisible.* Translated by Alphonso Lingis. Evanston, IL: Northwestern University Press.

Merleau-Ponty, M. 2003. *Nature: Course Notes from the College de France.* Translated by Robert Vallier. Evanston, IL: Northwestern University Press.

Mészáros, I. 1975. *Marx's Theory of Alienation*. London: Merlin Press.

Mills, S. 2011. "Concrete Software: Simondon's Mechanology and the Techno-social." *The Fibreculture Journal.* Available at http://eighteen.fibreculturejournal.org/2011/10/09/fcj-127-concrete-software-simondon%E2%80%99s-mechanology-and-the-techno-social/.

Mills, S. 2015. "Simondon and Big Data." *Platform: Journal of Media and Communication,* no. 6, pp. 59–72.

Mirowski, P. 2002. *Machine Dreams: Economics Becomes a Cyborg Science.* 1st ed. Cambridge: Cambridge University Press.
Mumford, S. 2006. "The Ungrounded Argument." *Synthese*, vol. 149, no. 3, pp. 471–89.
Mumford, S., and R. L. Anjum. 2011. *Getting Causes from Powers.* 1st ed. Oxford: Oxford University Press.
Pentland, A. 2012. "Reinventing Society in the Wake of Big Data: A Conversation with Alex (Sandy) Pentland." *Edge.* Available at http://www.edge.org/conversation/reinventing-society-in-the-wake-of-big-data.
Pentland, A. 2014. *Social Physics: How Good Ideas Spread—The Lessons From a New Science.* Melbourne: Scribe Publications.
Peters, J. D. 2015. *The Marvelous Clouds: Toward a Philosophy of Elemental Media.* Chicago: University of Chicago Press.
Plato. 1997. *Sophist.* In *Plato Complete Works*, edited by J. Cooper. Indianapolis: Hackett.
Prigogine, I. 1978. "Time, Structure and Fluctuations." *Science,* vol. 201, no. 4358, pp. 777–85.
Sartre, J.-P. 1940. *L'imaginaire.* Paris: Gallimard.
Shannon, C. E. 1948. "A Mathematical Theory of Communication." *Bell System Technical Journal,* vol. 27, July, pp. 379–423.
Siegert, B. 2013. "Cultural Techniques: Or the End of the Intellectual Postwar Era in German Media Theory." *Theory, Culture and Society,* vol. 30, no. 6, pp. 48–65.
Simondon, G. 1964. *L'individu et sa genese physico-biologique.* Paris: Presses Universitaires de France.
Simondon, G. 1980. *On the Mode of Existence of Technical Objects.* Translated by Ninian Mellamphy. London, ON: University of Western Ontario.
Simondon, G. 1989. *L'individuation psychique et collective.* Paris: Aubier.
Simondon, G. 1992. "The Genesis of the Individual." In *Incorporations*, edited by J. Crary and S. Kwinter, pp. 296–319. New York: Zone.
Simondon, G. 1995. *L'individu et sa genèse physico-biologique.* Grenoble: Éditions Jérôme Millon.
Simondon, G. 2001. *Du mode d'existence des objets techniques.* 4th ed. Paris: Aubier.
Simondon, G. 2008. *Imagination et invention.* Paris: Les Éditions de La Transparence.
Simondon, G. 2009. "Technical Mentality." Translated by Arne De Boever, *Parrhesia,* vol. 1, no. 7, pp. 17–27.
Simondon, G. 2010a. "The Limits of Human Progress: A Critical Study." Translated by Sean Cubitt, *Cultural Politics: An International Journal,* vol. 6, no. 2, p. 229.
Simondon, G. 2010b. "The Position of the Problem of Ontogenesis." Translated by Gregory Flanders, *Parrhesia,* vol. 1, no. 7, pp. 4–12.
Simondon, G. 2010c. *Communication et information: Cours et conferences.* Chatou: Les Éditions de La Transparence.
Simondon, G. 2012. "On Techno-Aesthetics." Translated by Arne De Boever, *Parrhesia,* vol. 1, no. 14, pp. 1–8.
Simondon, G. 2013. *L'individuation à la lumière des notions de forme et d'information.* 2nd ed. Grenoble: Éditions Jérôme Millon.
Simondon, G. 2014. *Sur la technique.* Paris: Presses Universitaires de France.
Simondon, G. 2015. "Culture and Technics." Translated by Olivia Lucca Fraser, *Radical Philosophy,* no. 189, pp. 17–23.
Stiegler, B. 1998. *Technics and Time, 1: The Fault of Epimetheus.* Translated by Richard Beardsworth and George Collins. Stanford, CA: Stanford University Press.
Stiegler, B. 2009. *Technics and Time, 2: Disorientation.* Translated by Stephen Barker. Stanford, CA: Stanford University Press.
Stiegler, B. 2010a. *For a New Critique of Political Economy.* Translated by Daniel Ross. Cambridge, UK: Polity Press.
Stiegler, B. 2010b. *Taking Care of Youth and the Generations.* Translated by Stephen Barker. Stanford, CA: Stanford University Press.
Stiegler, B. 2011. *Technics and Time, 3: Cinematic Time and the Question of Malaise.* Translated by Stephen Barker. Stanford, CA: Stanford University Press.
Terranova, T. 2004. *Network Culture: Politics for the Information Age.* London: Pluto Press.

Thomas, S. 2013. *Technobiophilia: Nature & Cyberspace.* London: Bloomsbury.
Toscano, A. 2012. "The Disparate: Ontology and Politics in Simondon." *Pli,* Special Volume.
Toscano, A. 2013. "Gaming the Plumbing: High-Frequency Trading and the Spaces of Capital." *Mute*. Available at http://www.metamute.org/editorial/articles/gaming-plumbing-high-frequency-trading-and-spaces-capital#sdfootnote19sym.
Turner, G. 1990. *British Cultural Studies: An Introduction.* London: Unwin Hyman.
Vanderbilt, Tom. 2009. "Data Center Overload." *The New York Times.* Available at http://www.nytimes.com/2009/06/14/magazine/14search-t.html.
von Foerster, H. 2003. *Understanding Understanding.* New York: Springer-Verlag.
von Neumann, J. 1966. "Computing Machines in General." In *Theory of Self-Reproducing Automata*, edited by A. W. Burks, pp. 31–41. Urbana: University of Illinois Press.
von Uexküll, J. 2010. *A Foray into the Worlds of Animals and Humans.* Translated by Joseph D. O'Neil. Minneapolis: Minnesota University Press.
Weaver, W. 1949. "Recent Contributions to The Mathematical Theory of Communication." In *The Mathematical Theory of Communication*, edited by C. Shannon E. and W. Weaver. Urbana: University of Illinois Press.
Wendling, A. 2009. *Karl Marx on Technology and Alienation.* New York: Palgrave Macmillan.
Whitehouse-Hart, J. 2014. *Psychosocial Explorations of Film and Television Viewing.* London: Palgrave.
Wiener, N. 1965. *Cybernetics or Control and Communication in the Animal and the Machine.* 1st ed. Cambridge, MA: MIT Press.
Wiener, N. 1989. *The Human Use of Human Beings: Cybernetics and Society.* London: Free Association Books.
Williams, R. 2001. "Culture Is Ordinary." In *The Raymond Williams Reader*, Oxford: Blackwell.
Williams, R. 2004. *Television: Technology and Cultural Form.* London: Routledge.
Winthrop-Young. G. 2011. *Kittler and the Media*. Cambridge, UK: Polity Press.
Zittrain, J. 2008. *The Future of the Internet—And How to Stop It.* New Haven, CT: Yale University Press.

Index

www.ingramcontent.com/pod-product-compliance
Ingram Content Group UK Ltd.
Pitfield, Milton Keynes, MK11 3LW, UK
UKHW041347230926
475560UK00013B/66

9 781783 481484